For more information:
quietriverpress.com

Practical Home Theater

A Guide to Video and Audio Systems

2017 EDITION

by Mark Fleischmann

QUIET RIVER PRESS

NEW YORK

ISBN 13: 978-1-932732-18-4
ISBN 10: 1-932732-18-7

This book is printed on acid-free paper.

quietriverpress.com

Dedicated to my mother, Lucille Fleischmann (1926-2012), who packed me into her pink-finned Plymouth and conveyed me to the Middlesex and Bound Brook, New Jersey public libraries when I was small and impressionable. This led to a fascination with books and a career pushing words in magazine, newspaper, online, and book publishing.

Thanks, Mom.

Table of Contents

Introduction

What is home theater?

I like home theater because it is lots of fun. Thank you, ladies and gentlemen, goodnight!

Welcome to home theater, square one. Though I fancy myself a home theater critic, credit for coining the term goes to Sam Runco, the projector king. I define home theater in three ever more specific ways: as a state of mind, as a room, and finally as a system playing in a room to a small, thrilled audience.

What is the home theater state of mind? It's the rush of emotional involvement that comes with a close-up shot on a big screen. It's the aesthetic thrill of hearing a rich orchestral score through an enveloping array of speakers. It's the quickening of the pulse that follows a gut-thumping subwoofer-borne explosion. It's about having more fun at home than you ever did at the movies. The movies have come home.

Where in the home does a home theater go? Wherever it can. Those with a spare bedroom or basement to devote to the hobby can end up with a luxurious screening room rivaling that of the early 20th century movie moguls. Then again, a home theater system may be a nonintrusive part of a multi-purpose room—until the lights go out. Then it transforms your living space, as that magic rectangle lights up and sound fills the room.

What are the elements of a *practical* home theater system? I'm talking about a movie *and* music system that takes best advantage of current technology but remains within a realistic budget. It starts with a big-screen display, coupled with a surround sound system. Movies, music, and other programming arrive via disc, DVR, streaming, download, computer, server, satellite, cable, telco, antenna, computer, tablet, smart TV, or smartphone—and fill in the past or future source of your choice. The right cables, remote, and other accessories tie the system together.

According to the Consumer Technology Association, a trade group representing manufacturers, about 41 percent of American households had home theater systems as of 2016—a figure unchanged since 2014—as compared with 6 percent in 1996. Given the universal love of movies and music, that still leaves some room for growth. Clearly many of us find the technology intimidating. The purpose of this book is to enable the reader to become familiar and comfortable with the technology behind home theater systems.

The book aims to be a comprehensive guide for beginners and intermediate-level readers. As a result, reading it from start to finish may be rough going (as my psychiatrist has noted rather pointedly). It's not designed to be something you'd take to the beach. However, it does work well as an answer book. Read the relevant chapter when you're about to buy something—the table of contents will send you to the right place. Or dip into the book when you have a question, using the index as your guide.

What is big-screen television?

A big picture is one of two key elements in a home theater system, in addition to surround sound.

Any of several kinds of big-screen television may form the basis of a home theater system. Which one is right for you is a question of room size as well as videophile taste.

For bedroom systems and small rooms in general, the best choice is a **flat panel** display using **LCD** technology.

For medium-sized rooms, larger LCDs are available, with **plasma** displays having been discontinued. **OLED**, a new display technology using an organic semiconductor situated between two electrodes, has replaced plasma in high-end sets and is growing.

For the biggest rooms, the biggest pictures come with two-piece **front-projection TVs** that splash the picture on a separate screen or even a blank white wall. Projectors use solid-state LCD or **DLP** technology to produce a bright, colorful picture.

Rear-projection TV is no longer an option, manufacturers having exited the category. **Direct-view** (or tube) TVs, even the digital ones, are obsolete and no longer good investments. Flat panels are more high-def-worthy and far more elegant.

Any of these video technologies has the potential to display **digital television (DTV)**, known in its most prevalent form as **high-definition television (HDTV)**. There is also an **ultra-high-definition television (UHDTV)** standard in the making, though the technology behind it is still formative. Whether it's a good investment right now is debatable—we'll get to that later. Please note that I now use the industry-sanctioned term **UHD** in lieu of the more informal **4K**, though they amount to the same thing.

The benefits of HDTV and UHDTV are best enjoyed on a large screen—the larger, the better. Big screens show off the detail and sharpness of HDTV and UHDTV *and* deliver greater emotional impact in general. While your eyes feast, your emotions get pulled into the story.

The "Television" chapter will discuss big-screen display technologies, and the basics of digital television, including resolution, screen shape, UHD, 3D, smart TV, tuners, and connections—finishing with advice on how to shop for a DTV.

What is surround sound?

Surround sound is the second key element of a home-theater system, after big-screen television.

A basic home surround system feeds a **5.1-channel** array of three front left/center/right speakers, two surround speakers on the side walls, and a subwoofer (the ".1"). In larger rooms, a **6.1-** or **7.1-channel** array adds one or two extra speakers against the back wall. New surround standards add two to four height speakers—so that a **5.1.4-channel** system, for example, would have five floor speakers, one subwoofer, and four height speakers. Each of the speakers in a 5.1 array specializes in a different part of the movie experience. The front-center speaker handles dialogue, while the other front and surround speakers provide ambience and effects, and the sub handles bass. To feed these speakers you would use either a one-box **surround receiver** or a two-box combination of **surround preamp-processor** and **multichannel power amp**.

Major movies produced over the past several decades have surround sound buried in their soundtracks. The surround formats that bring it home are licensed by Dolby Laboratories and DTS. There are roughly three generations of them.

The original **Dolby Surround** is most likely to be found embedded into old videocassettes. It is decoded in home gear by **Dolby Pro Logic**, **DPLII**, **DPLIIx**, or **DPLIIz**. The second generation, embedded into DVDs, is **Dolby Digital 5.1**, whose main competitor is **DTS 5.1**. There are also 6.1- and 7.1-channel versions, **Dolby Digital EX** and **DTS-ES**. The third and latest generation of Dolby formats includes **Dolby Digital Plus** and lossless **Dolby TrueHD**, while the third generation of DTS formats includes **DTS-HD High Resolution Audio** and lossless **DTS-HD Master Audio**. The lossless formats, especially DTS-HD MA, are dominant on Blu-ray discs. In streaming you are more likely to find Dolby Digital and Dolby Digital Plus, though DTS is starting to make inroads.

The incoming **Dolby Atmos** standard, an attempt to bring home the latest in cinematic surround technology, builds on the foundations of Dolby TrueHD and Dolby Digital Plus, but with an expanded speaker array, including height speakers. The DTS equivalent of it is **DTS:X**.

THX originally got started not as an alternative surround format but as a certification program that covers public movie theaters as well as every component in a home theater system including receivers, speakers, displays, cables, and software. Think of it as a high-end option that coexists with the Dolby and DTS surround standards. THX also offers listening modes including the useful **THX Loudness Plus**.

Audyssey licenses a variety of technologies including auto setup, room correction, low-volume listening modes, height/width listening modes, and bass-related modes. Some surround receiver manufacturers offer their own homegrown versions of auto setup and room correction technology to avoid the expense of using Audyssey.

The chapter on "Surround Sound" will tell everything you need to know about home theater speakers, receivers, preamp-processors, power amps, and the surround standards that infuse them with multichannel magic.

What's the cost?

Cost is an issue for readers of a book entitled *Practical Home Theater*. Regardless of what you buy, you may be about to lay out a lot of money for the pleasure of enjoying big-screen television and surround sound in your home. To make yourself an informed consumer, you will need to look at a broad selection of video and audio products—the best and brightest, as well as the cheapest and ugliest, and what's in between.

People are always asking me what they should buy. It may be tempting to have a purchase decision validated by an expert—that's the easy way out—but it's always better to understand what you're buying than to depend on some self-appointed guru. To get you started, I've named names in pictures and captions strewn throughout the book. Don't take these suggestions too literally. The products featured here are not necessarily *things you should run out and buy*. Rather, they are *things you should look at to become better informed*. Are the prices prohibitive? In many cases, yes, of course, but it's still helpful to learn what top-drawer products and critic's picks offer as a reference for evaluating other products. Only then can you gauge how much performance, and how many features, you need in your own system.

In the final analysis, how much should you spend for a home theater system? Other critics would pull dollar figures out of the air or recommend that you spend a certain proportion on speakers. Sorry, I refuse to play this game. But here are some tentative answers:

Don't spend more than you can afford. Don't spend enough to prevent you from buying, renting, or streaming movies and music to play on your system. Whatever you do spend should leave enough room for a good life, domestic harmony, and a nice bottle of wine with dinner. After dinner is when the fun begins.

Television

Welcome to the heart of home theater. Big-screen TV is half of the home theater equation. For some folks—not you blithe guilt-free consumers, but the remainder—buying a big-screen TV means breaking taboos. After all, isn't it the height of conspicuous consumption? Why buy a big expensive TV when stores are full of smaller and cheaper ones? And won't it take up an outrageous amount of room?

The answers: If you really want to enjoy movies at maximum emotional impact—and who doesn't?—a video display that serves as the centerpiece of a home theater system has to be large enough to dominate your senses. That's the only way to suspend disbelief and immerse yourself in the story. A big screen doesn't have to cost much, or take up much room. Wall-mounted flat-panel sets and front-projection systems can produce a huge picture with literally no footprint.

To get the best value out of a big-screen purchase, a consumer must become well-informed. Besides variations in screen size, digital TVs also vary in the way they handle different video formats, screen shape, and their need for outboard equipment to receive video from broadcast, cable, satellite, telco, internet, or other sources. Subsequent chapters will cover all these issues. But first, let's talk about big-screen display technologies.

Big screens

The introduction discussed various kinds of video displays in terms of scale: for small rooms, LCD TVs; for medium-sized rooms, bigger LCDs or OLEDs; and for the largest rooms, front-projection systems. This chapter will change focus from scale to display technology, though moving in roughly the same direction—from small to large.

UHDTV vs. HDTV vs. analog

Even if you forget everything else you read in this book, remember this bit of advice: *Don't buy an analog TV* (no, not even at a yard sale). The question is whether to buy a set based on mature high-definition television (HDTV) technology or evolving ultra-high-definition television (UHDTV) technology. In terms of model-years, HDTV has been around for close to two decades, whereas UHDTV is only into its first few model-years—and some aspects of it are still falling into place.

There's plenty of HD programming available via broadcast, cable, satellite, and internet. The best source is the Blu-ray disc, though high-def streaming is an alternative, with some compromise in picture quality. UHD programming is just getting started in Blu-ray and streaming. Cable and satellite companies are starting to roll out UHD services, some with picture-improving HDR technology.

So what are your fellow consumers actually doing? The vast majority of U.S. households now have HDTVs. While only one percent had UHDTVs in 2014, that figure was expected to climb to 10 percent by 2016, and UHDTVs are projected to take over at least half of the market by 2020. The practical home theater buff is one who buys a *good* digital display at a price the household can afford.

The sharper pictures of HDTV and UHDTV can bring several benefits. You may sit closer to the screen, enhancing the emotional impact of a good movie. Or you may keep viewing distance the same but have a larger screen. If the room's arrangement is settled, keep screen size and viewing distance the same, while enjoying higher performance in a screen the same size as your old TV. Whether you sit closer, increase screen size, or just enjoy higher performance, HDTV and

2

UHDTV make a difference in how well your home theater system delivers movies.

For more information on Ultra HDTV, see the chapter on "UHDTV and 3DTV."

Flat-panel display technologies

The most popular video displays are flat-panel sets using either LCD technology (for sets of any size) or OLED technology (for midsized to larger ones). LCD TVs in smaller sizes are now preferred to direct-view sets. Larger LCD and OLED panels have taken over from plasma and rear-projection sets. LCD, OLED, and DLP are referred to as **fixed pixel** displays because their grids of vertical and horizontal pixels are fixed. Antique CRT-based displays do not have fixed pixel grids.

The LG 65EF9500 ($6000) uses OLED technology, which has replaced plasma as the high-end display of choice. It has attracted strong reviews for picture quality, which is enhanced by HDR10 video technology.

Flat-panel displays: LCD

LCD stands for **liquid crystal display**. Flat-panel sets using LCD technology range from a couple of inches (in a smartphone) to 170 inches (the current record holder is a Samsung UHDTV). The best values are in the medium-sized 32- to 65-inch range though even larger sizes have become increasingly affordable. LCD TVs can be ultra-high-definition, high-definition, or standard-definition—make sure to buy the UHD or HD versions.

Most larger LCD sets are HDTV-worthy. They can sit on a pedestal and are usually light enough for wall mounting. Note that not all LCD TVs have interfaces for computer use and not all LCD computer monitors have video inputs for television formats (look for HDMI). However, mainstream TV makers now include broadband connectivity which supports a variety of networked media features. Many new companies—some of which have never before sold television sets in the United States—have enter the LCD TV market in recent years while traditional brands struggle to keep pace in an increasingly cutthroat marketplace.

How do LCDs work? Oblong liquid crystals serve as light valves when stimulated with current from tiny transistors printed on an underlying thin film. The light is supplied by a backlight behind the screen. Older sets used **fluorescent backlighting**, though current sets use **LED backlighting**, which allows a wider color spectrum, deeper black level, more accurate whites, more brightness, and lower power consumption.

There are two kinds of LED backlighting. **Edge-lit backlighting**, which lines the screen edges with LEDs, is dominant because it allows for slimmer panels at lower prices. However, the edge lights can't illuminate the screen evenly, leading to poor uniformity, which becomes apparent when the panel tries to reproduce a field of solid white or pastel color. Sets with even fewer LEDs are designed more for low pricing than for picture quality. The second type of LED backlighting is **full-array backlighting**. It places a field of LEDs behind the screen. This produces better uniformity and is helpful in larger screen sizes.

Edge-lit and full-array backlighting can both be enhanced by **local dimming**. In edge-lit sets, this targets specific groups of LEDs to improve contrast. In full-array sets, LED grids are dynamically manipulated from moment to moment, brightening or darkening sections of the image. The experts prefer this form of local dimming but it's limited to pricier high-end models.

In any form of LED-backlit set, there are fewer LEDs than pixels. That can cause light to spill over, causing halos around bright objects on dark backgrounds. OLEDS and plasmas aren't subject to this limitation.

The liquid crystals stand to attention when stimulated, then subside mechanically, and a bit slowly. This **response time** leads to a lag that causes motion-related distortions called **motion artifacts**. The motion artifacts peculiar to LCD TVs are often called **image lag** or **motion smear**. The best way to mitigate this problem is to increase refresh rate. This subject is discussed in more detail under "DTV by the numbers/Refresh rate shenanigans."

One major problem with LCD sets is that some do not support a wide **viewing angle**. Sit off to the side and dark-color reproduction diminishes. The whole picture takes on a purplish hue. This problem affects some brands more than others—watch for it before you buy.

Dead pixels are a manufacturing defect to which LCDs are prone. They are likely to appear right out of the box. When you uncrate your set, look for them. If you see too many, take advantage of your warranty and exchange the set immediately.

Watch for these defects when buying an LCD TV. But don't let them put you off LCDs entirely. Some are better than others. All share some major advantages. They can cost little, use less power than plasmas, and are lighter to mount. Look for Energy Star certification when buying any TV. You can look up certified models at energystar.gov.

Quantum dot: the latest in LCD

Quantum dot is not an entirely new display technology but rather a new kind of LCD display technology. Its goal is to improve color. It does this using a semiconductor crystal so small that it has the properties of both larger semiconductors and single molecules. At that point the properties of quantum physics come into play—hence the name quantum dot.

A TV image is created from a mix of three colors: red, green, and blue. In a quantum dot LCD set, a blue (as opposed to white) backlight creates blue and the quantum dots create red and green. The band gap between the two electrodes in each pixel can be precisely color-tuned. This process results in greater color saturation and brightness while wasting less light than a conventional LCD.

Unlike OLED, another nascent display technology, quantum dot sets are easy to manufacture. It's just a matter of swapping in the quantum dot film for another part. A quantum dot set can use any kind of LED backlighting—full array, edge lit, etc.—as long as it's blue. You're most likely to see it in Ultra HD sets though not for technical reasons. Manufacturers using quantum dot and labeling it that way include LG, Hisense, and TCL. Those using quantum dot under another name or no particular name include Samsung and Sony.

Quantum dot will benefit from HDR and other picture-enhancing technologies—see chapter on "UHDTV and 3DTV." Its relative ease of manufacture will enable it to spread further, faster, than pricier OLED.

Is OLED the future of flat-panel displays?

OLED stands for **Organic Light Emitting Diode**. OLED displays are electroluminescent—in other words, the pixels generate their own light, as in plasma. That provides a brighter and purer image than conventional LCD technology which relies on external backlighting. As plasma displays did, the current generation of OLED displays uses a heavy glass plate, but future generations are likely to use flexible polymers to provide bigger images, and less weight, at lower cost. In theory, an OLED display might be so flexible that you could roll it up.

Though OLEDs appear mainly in tiny displays on cell phones and music players, they are rapidly making inroads in TVs. Sony began selling 11-inch OLED TVs in 2007. Larger sizes soon followed from LG and Samsung. LG's 65- and 77-inch models control cost by using a novel **WRGB** technology that adds a white sub-pixel to the usual red, green, and blue.

A total of 500 OLED TV units were shipped in 2012 though the category is growing rapidly. Research firm IHS predicts that the OLED market will reach 830,000 units in 2016, 1.4 million in 2017, and 5.8 million in 2020. OLEDs allow manufacturers to claim higher prices in a market where flat panels have become commodity-priced.

Remembrance of things past:
direct-view, rear-projection, and plasma TV

Most major brands have exited the direct-view, rear-projection, and plasma TV categories in the North American market. The following section is therefore historical.

The industry refers to TVs with a single large picture tube as **direct view** to distinguish them from projectors that also use tubes. Literally speaking, you directly view the front of the tube, which has electrons flung at its surface from the narrow **yoke** at the back of the tube. HD-capable versions are available. Analog direct-view TVs are obsolete and digital versions are not far behind. Basic tube TVs can be as small as 5 inches or as big as 40, but anyone serious about home theater will step up to flat panels and projectors.

The shadow mask is what limits the resolution of direct-view HDTVs. As the holes in the mask are made smaller and more numerous, it becomes harder for light to get through them. So it's impossible for a direct-view HDTV to produce all the resolution of HDTV with all the light output consumers expect in a tube TV. No consumer-level direct-view TV can display all the scan lines in a 1080i HDTV test pattern.

The term **rear projection** formerly referred only to a one-piece set that uses a trio of cathode ray tubes to project an image from behind. However, **microdisplay**-based sets substituted either DLP technology licensed from Texas Instruments or liquid crystal panels. The term rear-projection is tricky—many front-projection TVs are able to throw an image from behind the screen. Mainstream RPTVs were boxy one-piece sets. RPTV and flat-panel sizes overlapped, but flat panels ultimately triumphed.

Rear-projection sets ranged in size from 40 to 70 inches, and occasionally up to 92, measured diagonally. Most models were in the fifties or sixties. In terms of dollars per square inch, they were the most economical HDTVs, but that wasn't enough to save them once flat panels dropped in price.

The recently deceased **plasma** category is deeply mourned by videophiles, some of whom still cling to their Pioneer Kuro and Panasonic models, which earned rave reviews in their heydays. Plasma came in sizes from 30 to 152 inches. **Gas plasma** technology produces light through tiny tubes filled with neon gas and coated with color phosphors. When activated by electric current, the gas produces ultraviolet light, which in turn energizes the cell's colored phosphor coating. The underlying plate is a thick glass sandwich, making plasmas the heaviest flat panels—mounting one is best left to a professional installer.

Most plasma displays had the full resolution necessary for HDTV though the category did not survive into the age of UHDTV. Their advantages over competing LCD technology included wider viewing angle, less motion smear, and superior reproduction of deep black and shadow detail—areas where LCD is still struggling to catch up. Because they did not rely on backlighting, they had better uniformity than LCDs, though they fared less well in rooms with direct sunlight.

If you're buying an old plasma display, you need to be aware of a few issues. One is **burn-in** or **image retention**. Like tube-based displays, a plasma panel can be damaged by static images that remain on-screen for long periods. Watch out for, among other things, video games and the bottom-screen crawl that accompanies some financial channels. To minimize the hazard, use moderate brightness and contrast settings (midpoint or lower) and control ambient light in the room. If burn-in does occur, run full-motion, full-screen material for 24 hours to mitigate the damage. Plasmas also behave oddly at **high elevations** where higher air pressure presses against the glass and causes a buzz. Some manufacturers adjusted sets that were shipped to such areas. **Longevity** was initially an issue—plasmas lose brightness as they age—but Panasonic models starting in 2008 take 100,000 hours to drop to half original brightness for 1080p models and 60,000 hours for 720p models.

LCD projectors

LCD competes head to head with DLP in projection. There are two kinds of LCD-based projectors: the **reflective** type, which reflects light off a liquid crystal panel, and the older **transmissive** type, which sends light through the panel.

Reflective LCD technology is usually called **LCoS (liquid crystal on silicon)**. Though slow out of the gate, it has matured to the point where some of the highest-performing projection sets use this technology, especially front-projectors from JVC and Sony.

The version of reflective LCD technology used by JVC is **D-ILA (direct-drive image light amplifier)**. At the heart of D-ILA is a tiny 1.22-inch liquid crystal panel about the size of a DLP chip. Because D-ILA reflects 93 percent of the light coming from the projector's lamp, it is highly efficient, and produces a very bright picture.

Sony's version of reflective LCD technology is **SXRD (Silicon X-tal Reflective Display)**. It uses three panels with what the manufacturer describes as a uniform liquid crystal cell gap, without any spacers in the image area, firmed up by an inorganic alignment layer. As a result the picture has a high pixel density while the space between pixels is as small as possible. And the process is robust enough for mass manufacturing at increasingly attractive prices.

D-ILA and SXRD can offer Ultra HD or 1080p Full HD resolution. The latest versions are simply stunning, with a deliciously fine-grained picture, plenty of brightness, and vivid color. If I were venturing into the market for front-projectors, I'd go for one of these.

DLP projectors

DLP was invented in 1987 by Dr. Larry Hornbeck of Texas Instruments. It produces an image by reflecting light off one or more chips containing thousands of tiny micro-mirrors. Texas Instruments used to refer to the underlying technology as **DMD** (**Digital Micro-mirror Device**)—now it's just called DLP technology. DLP projectors can be quite small, both because the chips themselves are small, and because the chips reflect heat as well as light, so the need to dissipate heat is minimized. Their reflective technology wastes little light, enabling them to produce a huge, bright picture. They are not as good at producing true black as tube and plasma displays. However, the pixels blend better than with other fixed-pixel displays, making the picture more seamless.

The latest DLP chips support UHDTV and 3DTV and are used in cinemas. However, don't assume all DLP projectors support the latest video technologies. Many DLP projectors, especially those for office use, may be sub-HD and therefore unsuitable for home theater.

A DLP projector may use one chip or three. All other things being equal, **three-chip DLP** models are more expensive—but better suited for high-end home theater or public theaters, which require extremely high brightness. Their optical innards filter light into video's three

The JVC DLA-X550R ($4000) uses D-ILA, the company's proprietary version of LCD projection technology. It is among the first JVC projectors to use HDR technology, a must for state-of-the-art Ultra HD.

primary colors (red, green, blue), routing each color to its own chip, then recombine the three light streams into one before the beam exits the lens at the front of the projector.

The more prevalent **one-chip DLP** models work differently, using a rotating **color wheel** to filter one color at a time, though it happens too quickly for the human eye to detect except in rare cases. Some eagle-eyed viewers claim to see a so-called **rainbow effect** that is most visible with bright objects rapidly moving against a dark background (such as a flashlight swinging in the dark). Newer models defeat the problem with much higher processing speed. Even on older models, it's usually so rare and fleeting that few people are bothered.

A color wheel must have at least three segments: red, green, and blue. That's a **three-part color wheel**. A **four-part color wheel** adds a colorless segment to increase brightness. In a **six-part color wheel**, the red, green, and blue segments are doubled, reducing the rainbow effect to (in my opinion) virtual nil. Mitsubishi's version of the six-part wheel uses red, green, and blue plus yellow, cyan, and magenta. This brings a huge increase in color fidelity—daffodils actually look yellow.

But the best color wheel may be no color wheel. DLP projectors can use flashing **LEDs** (light-emitting diodes) to replace the color wheel and the arc lamp that goes with it. These lamp-free projectors operate well even in rooms with lots of ambient light. Among the other benefits are no rainbow effect, longer life, and more uniform performance over the lifetime of the projector.

Traditional CRT projectors

In bygone days, the video display of choice for high-end videophiles was the **CRT**-based projection TV. CRT stands for **cathode ray tube** and a CRT projector uses three of those big, heavy tubes in video's primary colors of red, green, and blue to produce an image. The best CRT-based projectors can accurately display just about any video format that exists. Their performance in some areas, such as black level and shadow detail, remains unsurpassed.

CRT projectors made more demands on the consumer than newer display technologies. They were larger, heavier, and more difficult to mount than DLP or LCD projectors. The best CRT projectors were the largest ones with nine-inch (or larger) tubes. More affordable models with seven-inch tubes could also produce a beautiful picture but didn't offer as much resolution. (A nine-inch CRT projector could display all the scanning lines in a 1080i HDTV picture; a seven-inch CRT projector usually could not.) For best performance, surviving CRT projectors require professional installation. They also need regular adjustment (called **convergence**) to keep all three tubes in alignment. Otherwise, color fringing will result, and will worsen over time. In addition, because green color phosphors age more rapidly than red and blue, a technician must occasionally adjust the **color balance**.

Though CRT projectors have the potential to produce a fantastic picture, they do have a couple of performance disadvantages. One is **blooming**—when the tubes are overdriven, images get bent out of shape. Another is burn-in, or image retention—static images from videogames and other graphics-heavy material may leave a permanent imprint on the tubes. This is also a problem with plasma but it's a bigger one with tubes.

If you buy a used tube projector, make sure it has been **retubed**. Replacing the CRTs adds to the cost but ensures that you'll get the full benefit of your investment.

Projection screens

The most significant accessory for a front projector is a screen. Choosing the wrong screen can negate your investment in a good projector. The screen directly affects how you'll perceive the projector's performance (remember that when taking in projector demos!). Consider it part of the projection system's cost.

Gain, or the amount of light the screen reflects, is the key spec. Screens with a gain of 1.0 reflect back precisely as much light as they receive from the projector. Those with a gain of more than 1.0 offer a tradeoff. They provide more brightness to viewers sitting at the center but less to viewers sitting at the sides. Therefore high-gain screens provide more brightness, but less uniform brightness, a fact reflected in another specification, **viewing angle**.

Avoid extremely high-gain screens if you have a large family, or plan to host screening parties with many guests. After all, you want audience members at the sides to see a decent picture. A good limit would be a gain of 1.3—for instance, the Stewart Studiotek 130 (developed by the influential video expert Joe Kane) or the Da-Lite Cinema Vision. On the other hand, if your home theater is long and narrow, or you do most of your viewing alone, the extra brightness afforded by a high-gain screen can do a lot to enhance the brightness of a projector. A high-gain screen also allows some projectors to work better in the presence of ambient room light—though reducing the light level in the room is a better solution for most than using a high-gain screen. High-gain screens are especially suitable for 3D due to the latter's reduction in brightness.

With fixed-pixel projectors now dominant, screens designed to compensate for their shortcomings have become a necessary accessory. These screens are literally grey, and that enables them to get a deeper black level and more shadow detail out of the projector. The Stewart FireHawk's gain is slightly greater than 1.0 but some other DLP-friendly screens have lower gain. Examples include Stewart's GrayHawk, with a gain of 0.95, and Da-Lite's High Contrast Da-Mat, with 0.8.

Though the following statement may attract derision from high-end videophiles, you may not need a screen at all if you're using an LCD or DLP projector. A clean wall surface painted a neutral white can serve as a stealth screen. This is not the best way to maximize the performance of a projector—but it may be an attractive option for someone who does not want to call attention to a home theater system when it is not in use. Another advantage of using a wall is that you can zoom the image in and out without having to make any binding decisions about how large it should be.

One of the advantages of using a screen is that it will have **black masking** around the edges. Masking helps frame the image and, by providing a contrasting element, makes it subjectively appear brighter and more vivid. Masking can be made adjustable to fit both widescreen and nonwidescreen **aspect ratios** (screen shapes).

Screens not mounted within frames can curl, and bend shapes on-screen. **Framed** or **tensioned** screens don't have this problem; they can sit on legs or their frames can be mounted to the wall. Don't want to see your screen when there's no show? Look at **roll-up** screens with motor-ized mechanisms that withdraw the screen when it's not needed.

Video processors

Because front-projectors produce such a huge picture, they merci-lessly magnify the flaws in standard-definition and analog signals. Even if you are a digital video enthusiast, this is still an issue if you want to watch DVD, VHS tapes, laserdiscs, or other analog video sources. If you invest in a UHD projector, you'll also be watching a lot of upscaled 1080p material, since there is currently little native UHD programming available.

To deal with this problem, some projectors function better with the help of **video processors** or **scalers**. Nowadays they're usually built into the projector (or a surround receiver or surround pre-pro) but they can also be bought as separate devices. They increase the number—and therefore reduce the size—of scan lines on the screen. Increasing the number of scan lines does not increase the amount of information in the picture, but does rearrange the existing picture information to better advantage, making the line structure less visually intrusive, and producing an image that more closely approximates the feel of film.

The crudest video processors are merely **line doublers**; **line quadruplers** or **line interpolators** do a more thorough job. Most consumer-level projectors already have built-in video processors. The question, then, is whether an outboard processor would improve on the built-in one's performance. Going from a simple line doubler to a more sophisticated video processor can produce a visible difference. A high-end video processor used to be a five-figure investment. However, with a progressive-scan disc player and/or a video processor costing a few hundred dollars, you can get performance that would have cost tens of thousands just a few years ago.

To get the benefit of a video processor, your projector will need to generate the scan lines that produce the picture at or above a certain speed. It should have a **scanning rate** of at least 31.5 kHz (kilohertz). That's the same scanning rate required by 1080i HDTV; 720p requires 45 kHz. A line quadrupler requires a projector with a scanning rate of 63 kHz and up. For HD 1080p you'll need 67.6 kHz. For UHD you'll need 110 kHz. However, there's more to mating projectors and processors than just keeping an eye on the scanning rate. A projector may have an optimum scanning rate, below its maximum rate, that maximizes light output and sharpness.

The increasing popularity of fixed-pixel displays that convert all signals to a single scan/refresh rate has led to a new category of video processor known as the **single-rate scaler**. Why invest in a sophisticated multi-format scaler when all you need is one that handles a single format and does it well? Some single-rate scalers are offered as accessories with matching projectors.

Warning: Don't go it alone! Speak to your projector's manufacturer, as well as your local high-end video dealer, about what kind of video processor (if any) is appropriate for your projector. Take your time and watch a few demonstrations—with *your* model of projector—and observe differences before and after processing.

13

DTV by the numbers

To get a sense of what HD and UHD really mean, it helps to know something about resolution.

Understanding resolution

- resolution, native vs. measured
- resolution, vertical vs. horizontal
- lines of horizontal resolution

The spec known as **resolution** determines both **detail** (the amount of information in the picture) and **sharpness** (the amount of brightness or contrast at the edges of objects). It is the one of most important specs to look at when you're buying a TV and one of the most potentially misleading.

There are two kinds of resolution: native and measured. **Native resolution** indicates the maximum number of visible **pixels** (**picture elements**) that a video display is capable of delivering, as dictated by its physical structure and underlying electronics. **Measured resolution** indicates performance as measured on a test pattern. These two kinds of resolution may be interrelated but are not necessarily identical. Measured resolution may fall short of native resolution, but can never exceed it.

In digital television, native resolution is a straightforward pixel grid arranged in vertical columns and horizontal rows. For example, if a projector has **vertical resolution** of 1080 pixels and **horizontal resolution** of 1920 pixels, on a spec sheet, that would read "1080 x 1920" (or vice versa). Note that *the words vertical and horizontal refer to the method of counting, not the things being counted.* Vertical resolution describes horizontal rows of pixels, counted vertically, from the top of the screen to the bottom. Horizontal resolution describes vertical columns of pixels, counted horizontally, from side to side. (Because the pixels may not be placed precisely on top of one another, horizontal resolution may also be counted as the number of pixels on each horizontal line.)

In obsolete analog television, resolution was measured differently—as **lines of horizontal resolution** or, more simply, **TV lines** (**TVL**). Do

14

not confuse this spec with the number of horizontal scan lines. It refers to the number of visible vertical lines that can be counted horizontally—not across the full width of the screen, but across a horizontal distance no greater than the picture height. If you visualize it correctly, you'll see a square made up of alternating black and white vertical lines.

How much resolution is enough? In the age of HDTV, you need at least 720 by 1280 pixels, Full HD is 1080 by 1920, and Ultra HD is 2160 by 3840. We'll discuss DTV formats in the next section.

Blu-ray test discs allow you to test resolution and adjust picture settings on any set linked to a disc player. Note that DVDs are standard-, not high-definition. With those, you can measure lines of horizontal resolution only up to the original DVD format's limit of 540 TV lines (or less). So get the Blu-ray versions—for example, *Digital Video Essentials: HD Basics* (now available in an updated Ultra HD release) or the *Spears & Munsil HD Benchmark*. Take one of them to the store and run your own tests if you really want to see how TVs perform. (Don't be surprised if you get thrown out of the store.)

Meet HDTV, EDTV, and SDTV—and introducing UHDTV

Before we get into digital television, let's note what it replaced. The now-obsolete North American **analog television standard** is named **NTSC**, after the **National Television System Committee**, which finalized it in the early 1950s. It has a total of 525 scanning lines. That determines its vertical resolution, though the two numbers are not identical. That is because not all of the scan lines are **active** (meaning visible)—some are used for control purposes, to stabilize the picture, or to deliver closed captioning. Therefore the measurable and visible vertical resolution of an NTSC set is about 480 lines, with each horizontal line holding about 440 pixels.

Eighteen different formats go into the **digital television (DTV)** or **ATSC 1.0** standard as defined by the **Advanced Television Systems Committee (ATSC)**, the standard-setting body for which the DTV standard is officially named. However, they fall neatly into three groups. **High-definition television (HDTV)** offers a picture with roughly twice as much information as analog TV. **Enhanced-definition television (EDTV)** offers a picture just slightly sharper and more detailed, but much cleaner, than analog, thanks to its use of progressive scanning (see below). And finally, **standard-definition television (SDTV)** is a lower-resolution format offering modest advantages over analog TV.

15

Note that the SDTV designation once referred to the EDTV format—older product literature may refer to EDTV as SDTV. Also that **Ultra HD** isn't part of the broadcast standard (though the ATSC folks are working on version 3.0). And the fun has just begun!

Additional versions of the ATSC broadcast standard are following the original version, **ATSC 1.0**, described above. **ATSC 2.0** was intended to support IP (internet delivered) video as well as better video compression, program guides, video on demand, and other features. But development skipped ahead to **ATSC 3.0**, which will bring TV broadcasting into the era of Ultra HD and height-enhanced surround. As this edition was being put to bed in late 2016, ATSC 3.0 hadn't made it past test broadcasts, and the technology was still formative.

Following are two tables that'll help you better understand digital television technology. The first one sums up all 18 of the video compression formats sanctioned by ATSC 1.0. These are the formats in which DTV may be transmitted over the air (satellite and cable transmissions don't always conform to these numbers). The second table describes the formats used in some DTV sets.

Vertical lines	Pixels	Aspect ratio	Picture rate
1080	1920	16:9	60i, 30p, 24p
720	1280	16:9	60p, 30p, 24p
480	704	16:9 or 4:3	60p, 60i, 30p, 24p
480	640	4:3	60p, 60i, 30p, 24p

The first column above represents the number of horizontal scanning lines that make up the picture and determine its **vertical resolution**. In the second column is the number of pixels per scanning line, which determines **horizontal resolution**. **Aspect ratio** is screen shape (16:9 is widescreen, 4:3 a conventional screen). As you glance down the far righthand column, you'll notice that the number of pictures transmitted per second is variable, and that pictures may be transmitted as full **frames** (**p** is for progressive scanning) or as half-frames, known as **fields** (**i** is for interlaced scanning). The distinction between progressive and interlaced scanning is discussed further a few pages down.

Format	Resolution	Aspect ratio	Picture rate
4K Ultra HD	2160 x 3840	16:9	60p, 30p, 24p
1080p HDTV	1080 x 1920	16:9	60p
1080i HDTV	1080 x 1920	16:9 or 4:3	60i
720p HDTV	720 x 1280	16:9	60p
480p EDTV	480 x 848	16:9	60p
480i SDTV	480 x 640	4:3	60i

This second table will interest the DTV shopper more—these formats are not for transmission, but for actual display on a DTV you may be considering. Some are direct implementations of ATSC formats; others are conversions to a display's native format.

Pride of place at top goes to the new **4K** (or **Ultra HD**) format. UHDTVs are now available. Programming is starting to roll out, with UHD Blu-ray and streaming, and satellite operators and Comcast expected to follow. Aside from these sources, much of what you watch on a UHDTV would be upscaled from lower resolutions.

Next is the 1080p format, also known under the marketing term **Full HD**. Like UHD, it does not exist in the ATSC 1.0 specs—in other words, there is no 1080p broadcast programming—but many new HDTVs have 1080p signal processing. This is mainly an upconversion format. When the set receives any kind of signal, it converts to 1080p. This may provide a more seamless picture, if done well, but doesn't really afford more picture information. 1080p is established in Blu-ray at the highest level of quality and in streaming at lesser levels of quality.

Next is the 1080i format used in some HDTV broadcasts. It provides the maximum number of scan lines in an interlaced format. (See the next section on "Interlaced vs. progressive scanning.")

The last three items describe lower-priced displays, high-definition and enhanced-definition. One of them provides the same resolution as the 720p ATSC format with progressive scanning. Others provide the same 480 vertical pixels as the ATSC's EDTV format but stretch the horizontal pixels out to 848 to optimize the imaging chip for widescreen use. At the bottom is 480i SDTV, the closest digital equivalent to the now-gone analog broadcast standard.

At the far right, most of the numbers have all suddenly standardized at 60 fields or frames per second. That's to minimize flicker, which would occur at lower **screen refresh** rates. Note, however, that the formats delivering 60 full frames per second have a major advantage over those delivering only 60 fields (half-frames) per second.

These are just some of the ways in which DTV may be implemented. You'll sometimes see pixel-grid numbers other than these.

Interlaced vs. progressive scanning

When the video salesperson mentions 1080p, what do those numbers and letters mean? The numbers are easy—they're the scan lines and pixels that produce the picture and determine its vertical resolution. The letters, on the other hand, refer to interlaced and progressive scanning, the two different methods that analog TVs and computer monitors use to produce a picture. A DTV set may use either method.

Interlaced scanning uses two alternating **fields** of horizontal scan lines to form a **frame**—that is, a full still picture. First, one field worth of lines traces a ghostly half-picture from top to bottom, with spaces in between each line. Then a second field comes along and fills in the gaps, interlacing with the first.

Progressive scanning is simpler. It is not measured in fields—only in frames, as the horizontal scan lines trace a full frame from top to bottom. Thus is created, frame by frame, the illusion of moving pictures. This is how films have been made from Sir Charles Chaplin's two-reelers to the latest blockbuster. Most DTVs now use progressive scanning.

Both formats entered the DTV standard as a compromise between TV makers and broadcasters (who favored interlacing for its bandwidth efficiency) and computer makers (who favored progressive). Interlaced scanning is twice as efficient because a field contains only half as much information as a frame. However, 1080i survives mainly as a broadcast format. Most digital TVs now use some form of progressive scanning.

1080i vs. 720p

HDTV comes in two broadcast formats. Each one does something better than the other. Both are considered to be HDTV by the Advanced Television Systems Committee (ATSC), which designated the formats following years of research and debate, and by the Consumer Technology Association, an industry trade group.

The **1080i** format produces a picture with 1080 horizontal scan lines, in two halves, interlaced together (as explained above). The **720p** format produces a picture with 720 lines, one full frame at a time, with progressive scanning—no interlacing.

18

The key distinction between 1080i and 720p is spatial versus temporal resolution. Better **spatial resolution** is the strength of 1080i. Because it uses more scan lines per frame, it produces a sharper and more detailed picture when the image is frozen or barely moving. Stronger **temporal resolution** is the advantage of 720p. It can reproduce rapidly moving objects with less blurring, as well as cleaner graphics, because it uses progressive scanning that leaves the frame intact. The 1080p format, delivered by Blu-ray, combines the strengths of 1080i and 720p—as we'll see in the next section.

Except for PBS, whose affiliates are free to use either 1080i or 720p, the major broadcasters have adopted one format or the other. CBS and NBC are firmly in the 1080i camp. ABC, after initially rebelling against the whole notion of HDTV, has become a stalwart supporter of 720p. Fox was initially the least enthusiastic about HDTV in general, electing to deliver widescreen 480p instead—still a big improvement over analog—but eventually went 720p, or 1080i in some of its cable channels. That doesn't mean you have to buy a separate DTV for each format—any DTV should display pictures in any format by upconverting or downconverting to its native resolution (see next section).

1080p: Full HD

HDTV in a **1080p** format, also known as **Full HD**, has moved from pro/archival use to consumer use. It combines the strengths of 1080i and 720p, producing a 1080-line picture in full frames.

Most HDTVs support 1080p as an upconversion format—in other words, they display other signal formats in 1080p to gain a small edge in performance. However, true 1080p *signals* were initially limited to industrial users. Among other problems, they are too large to fit onto a DVD or through a broadcast channel, and therefore were offered only as a display format for upconverted signals.

The advent of the Blu-ray disc format has brought 1080p to consumers. You won't get it off the air but at least you can get it (in order of quality) via disc, telco-fiber, cable, satellite, or streaming.

Until recently, 1080p was the state of the art in mass-market HDTV. In fact, it may exceed the limits of human vision. Even if your vision is perfect, you can't see the pixels on a 720p display beyond a certain distance—so whether you really need 1080p (let alone UHDTV) is a function of viewing distance and screen size. For smaller HDTVs at correct viewing distances—not to mention the many 1080p smartphones—

1080p may be overkill (though it certainly doesn't hurt).

For information on UHDTV, skip two chapters ahead to "UHDTV and 3DTV."

24 fps: the movie lover's frame rate

In addition to 1080p, another bit of alphanumeric magic you'll find on spec sheets is **24 fps**. It indicates that movie-based material can pass from some high-def signal sources (like Blu-ray) to a compatible set at 24 frames per second. This is the same frame rate movie projectors use. It's rather flickery, though, and that's why movie theaters double the frame rate from 24 to 48. In television, the frame rate is converted from 24 to either 30 or 60, a process called 3/2 pulldown (see below). How well your HDTV, disc player, or other signal source performs 3/2 pulldown matters a lot more than whether the source is 24 fps. Note that 24 fps and 3/2 pulldown apply only to film-based material, not programming shot on video.

Refresh rate shenanigans

When shopping for an LCD TV, look for the highest **refresh rate** you can find to minimize the motion artifacts brought on by the slow response times of liquid crystals. This is a problem with LCD sets, not with OLED or other sets.

Many models have a **120 Hertz refresh rate**, which means they multiply the frame rate of incoming signals to 120 frames per second. Some go to refresh rates of **240 Hertz** or greater. This can look good, but only if the underlying circuitry keeps up. In public demos, this has been shown to reduce the blurring that occurs when the camera pans across something—say a face, or a newspaper. However, in demos not controlled by manufacturers, the difference is often much more subtle.

The increased refresh rate may not account for all of the difference. That's because it's often combined with **smoothing** or **anti-judder** circuitry, which is sometimes separately adjustable. This produces a more solid, though not necessarily more film-like, image. Some prefer it off.

Manufacturers achieve higher refresh rate numbers in various ways. The best method is **frame interpolation**, which creates new frames as composites of existing ones. This works well for programs shot on video though not so well for those shot on film, and some videophiles like to turn it off when viewing a movie. But some manufacturers enhance sharpness using **backlight flashing**, turning off the backlight between frames. Another version of this is **backlight scanning**, also known as **black frame insertion**, which dims the backlight in sections.

The upshot is that one so-called 120 Hertz set may be a 60 Hz panel with backlight flashing, resulting in loss of light output; while another may be true 120 Hz with frame interpolation. Manufacturers use fancy names to obfuscate this feature and spec sheets are becoming increasingly deceptive. *c|net* now makes it a point to specify a panel's true underlying refresh rate in product reviews.

With Ultra HD, it is more important than ever to find out the panel's true refresh rate. The panels of most UHDTVs are 60 or 120 Hz, regardless of what is claimed.

Back in the plasma era, some manufacturers specified a **600 Hz refresh rate**. This was unnecessary because plasmas already had better motion reproduction than LCDs. But marketers were afraid consumers would see high refresh rates specified for LCDs and then assume plasmas are inferior. So they talked up the 600 Hz refresh rate. It was technically correct because plasma pixels flashed on and off at high rates. But all plasmas worked this way, so the 600 Hz refresh-rate spec was meaningless.

Upconversion and downconversion

A key feature of any HDTV or UHDTV is its ability to convert video from other formats to the **native resolution** that best fits the set itself. This is especially true of UHD, since there is relatively little UHD programming. UHDTV owners watch upconverted HD and other programming most of the time because the existing broadcast, cable, satellite, and telco delivery systems are limited to 1080i, 1080p, and other sub-UHD formats.

Every display is physically and electronically optimized to work best with a certain number of pixels or scan lines, which determines detail and sharpness. For instance, most HDTVs have a native vertical resolution of 1080p, meaning they display a picture with 1080 scan lines in one pass per frame.

What happens when a 1080p set gets something other than a 1080p signal? After all, there's no such thing as 1080p on the air—HDTV broadcasts are either 1080i or 720p—and some channels or programs may operate at lower resolution. When a native 1080p set receives, say, a 480p signal from a DVD player, or analog content from a legacy format, it may **upconvert** the signal to 1080p.

That doesn't mean the set turns fuzzy signals into sharp ones. It just means the set multiplies the 480 lines to 1080. There may be 1080 scanning lines onscreen but they display 480 lines worth of information. This is analogous to what happens when you play a lousy MP3 encoded at a paltry 128 kbps through a high-end stereo system. It still sounds lousy.

By the same token, an EDTV set has to **downconvert** higher-resolution 1080i signals to its own native 480p, which is like putting a CD in a boombox. The boombox will still sound like a boombox though the CD will allow it to sound as good as a boombox can sound.

Note that sets that convert film-based programming from 480i to other native resolutions include **3/2 pulldown** to compensate for the mayhem that occurs when 24-frames-per-second film is converted to 30-frames-per-second video. As an alternative, 3/2 pulldown can also be performed by a progressive-scan disc player. For more about this topic see "Disc Players/Progressive-scan DVD-Video players."

Conversion determines how your DTV will perform with different sources of programming and you should consider every possible angle.

Interpreting TV specs

- video S/N
- chroma AM S/N
- chroma PM S/N
- brightness and lumens
- color temperature

Resolution is not the only way of quantifying a display's performance. Here are some more specifications. Note that this discussion is about TV specs, not TV controls. The controls don't dictate the specifications. They merely adjust the set's performance within the limits of its specs.

Expressed in decibels (dB), **video S/N (signal-to-noise) ratio** is another key spec. It describes the amount of noise, snow, or other visual contamination in the brightness portion of the video signal. The **chroma S/N ratios** do the same for the color portion, which is separate. There are actually two kinds of chroma S/N. **Chroma AM** (amplitude modulation) describes how much picture noise affects saturation or intensity of color. **Chroma PM** (phase modulation) describes how much noise affects hue, or color fidelity (a fire engine should not be purple). If you want a clean picture, pay attention to these less often quoted specs.

Brightness—again, the spec, not the picture control—is measured in the whitest part of the picture with the contrast control at its maximum setting. (And by the way, only a fool would actually watch a movie that way.) The spec may be given in two ways. A projector's light output is given in **lumens** or **ANSI lumens** (per the American National Standards Institute). ANSI lumens are measured differently than regular lumens. Using the ANSI specification method, numbers tend to be dramatically lower. Reflected light—what you actually see on the screen—is measured in **footlamberts** (**ftL**). The broadcast industry standard spec for professional video monitors is about 30 ftL. Spec inflation is rampant here because what is feasible for measuring is inappropriate for normal viewing. With projectors, less than half of the 30 ftL figure is normal, because they are designed to be used in darkened rooms.

Don't get hung up on these numbers. Instead, as you eyeball projectors in showrooms, try to get the picture projected to the same size that your own screen will be (brightness drops as picture size increases). And ask the dealer to adjust the light level to approximate what's going on at home. Keep in mind that projection lamps lose light output as they age, so neither the specs nor the showroom demo will reveal everything about the longterm experience of using the projector in your home.

Many videophiles think the correct **color temperature** (or color balance) is indispensable to achieving a truly filmlike image. Measured in **Kelvins** (some prefer to say "degrees Kelvin"), color temp should be precisely 6500 according to the television standard. That's neutral white, and therefore a good place to start if you want accurate color. Unfortunately, most sets sold today are not shipped in a mode that meets the standard because manufacturers think turning the whites blue, and thus cranking up brightness levels, will make their TVs more appealing under bright showroom lights. Over time the effect looks more and more artificial as the viewer realizes that all other colors have been skewed to generate a bluer, brighter white. Most sets offer the option of choosing 6500 Kelvins from among several settings (often called the **movie mode** or **film mode**). Others can be set to the standard by a technician, such as those certified by the Imaging Science Foundation (visit imagingscience.com for more details). Finally, note that color temp may not be fully **linear**, instead varying according to the level of brightness. But setting it as close to the standard as possible will bring an enormous improvement in color fidelity.

23

The shape of DTV

Television has changed its shape. HDTV has brought greater resolution than analog TV, but that is not its only benefit. Just as important, but less heralded, is the transition from old-style narrow television to the wider screens of today.

Unfortunately, your local retailer may not yet be widescreen-hip. Visit a chain store selling HDTVs and you're likely to see a shocking display of technical illiteracy: the stretching of nonwidescreen signals on widescreen displays. There is something almost obscene about a video image that is stretched out of shape. Please, don't try this at home!

Screen shape, 16:9 vs. 4:3

TV programming comes in two screen shapes, also known as **aspect ratios**. Widescreen shows have a **16:9** ratio of screen width to screen height, similar to that used by most contemporary filmmakers and HDTV programs. Older shows mimic the **4:3** ratio of older movies and analog television. (You may read those numbers as "sixteen by nine" and "four by three.") To reduce those numbers to common denominators, that's 16:9 vs. 12:9, or 1.78:1 vs. 1.33:1, so widescreen proportions are about a third wider.

Until the 1950s, all movies and all television programs were made in a ratio of 4:3. Films were first to break the mold. Mid-20th-century movie moguls thought widescreen film would be a good stick with which to beat television. They figured audiences would flock to see films made in a wider aspect ratio.

With such films as *Ben Hur* (1959) and *Spartacus* (1960), both of which won Oscars for their (literally) spectacular cinematography, Hollywood redefined action movies and ushered in the age of widescreen viewing. These films were made in CinemaScope, which boasted a super-wide aspect ratio of 2.35:1.

Widescreen moviemaking didn't adhere to a strict standard. Some filmmakers preferred a less extreme width. Some even designed films to be viewed in any of several aspect ratios. A good example would be Alfred Hitchcock's *North by Northwest* (1959), which the Master of Suspense intended to be viewed at 2:1, 1.33:1, or anything in between.

Today, most filmmakers favor an aspect ratio of 1.85:1, though some prefer a wider 2.40:1. The former is close to the 1.78:1 (16:9) aspect ratio adopted for widescreen formats in the DTV standard. HD television programming matches the standard precisely.

Panning-and-scanning

Trimming widescreen movies for old-style 4:3 screens resulted in cinematic butchery. Today's 16:9 screens are more accommodating to widescreen movies, though some movies are even wider than that. They are adapted to 16:9 with letterboxing (see below) though they may get cropped during mastering or stretched by the misuse of TV aspect ratio modes at home.

Typically, it was the width—not the height—that was trimmed when a widescreen movie was adapted to the 4:3 TV screen. The traditional means adopted for this cinematic surgery was **panning-and-scanning**, which allowed a technician to choose which side of the screen should be cropped the most. This is the technique used to transfer most films to video for VHS and other legacy analog media. Panned-and-scanned DVD releases are labeled **full screen** (as opposed to widescreen) though they don't display the full image.

Watching panned-and-scanned video can be disconcerting. Ever watch a scene with two people talking face to face, but the backs of their heads are cut off, with a veritable sea of space between their truncated faces? Or maybe one is cut out entirely, appearing to yap at the side of the frame? Or maybe the camera seems to be weaving drunkenly to and fro between them? Ever see part of a movie that just doesn't make sense—like a character reacting to something outside the frame that you can't see? Do camera moves in general appear clumsy, as though the camera operator were having a really bad day?

Those are all the effects of butchering widescreen films with the pan-and-scan technique. In response, directors and cinematographers started working defensively, visualizing a 4:3 sweet spot as they composed the widescreen image, and making sure nothing significant fell outside it. This limited their use of the widescreen medium. Today 16:9 is the HDTV standard, so filmmakers can shoot widescreen material for widescreen sets. However, many movies are shot in aspect ratios wider than 16:9, and that must weigh on the mind of a director—wondering how much of the image may be cropped for video release.

25

Letterboxing

As an alternative to panning-and-scanning, the video industry of the 1980s came up with **letterboxing**. This technique fits a widescreen image onto a narrower screen by placing blank (black or grey) bars at the top and bottom of the screen, thus allowing the full width to be viewed at the expense of height. Letterboxing was a popular stylistic gimmick among music-video directors of the '80s. It was a good way to adapt widescreen movies to old 4:3 TV screens and is still a good way to adapt wider-than-16:9 movies to 16:9 screens.

Letterboxing has drawbacks, reducing picture height and resolution. It fails to take advantage of a screen's total height. It also reduces resolution because fewer of the scan lines that produce the picture are devoted to the image itself. Some are used to record the blank letterbox bars, which sort of turns the TV into a six-shooter loaded with four bullets and two blanks. Most problematic of all, letterboxing causes uneven burn-in of plasma panels.

These problems haven't prevented letterboxing from proliferating in several formats. Letterboxed laserdiscs used to be quite common. For many years they were the format of choice for videophiles. Even letterboxed DVD and VHS releases are not unknown, though further reducing the already poor resolution of VHS is not necessarily a good idea (don't even try reading the end credits). Blu-ray and DVD use letterboxing on the sides for pre-widescreen 4:3 movies, and on the top/bottom for contemporary movies wider than 16:9.

On a widescreen DTV fed with some older letterboxed material, the blank bars appear not only at top and bottom, but on all sides. Selecting the set's letterboxing (or zoom) mode will enable the image to fill the screen but will not counteract the reduced resolution.

A 16:9 TV fed with 4:3 material will have bars at the sides. To avoid this, some older TV programs (for example, *Seinfeld*) are now cropped, losing part of the picture at the top and bottom.

Anamorphic: a better mousetrap

An alternative to old-style letterboxing was found for DVD. It's called **anamorphic** video and it's the norm in the majority of widescreen DVD movie releases. (Blu-ray, like HDTV widescreen formats in general, is not anamorphic.) All consumer-level DTVs support anamorphic video.

How does anamorphic video work? To oversimplify a little, it squeezes and unsqueezes the picture to achieve maximum sharpness. Maybe that suggests the picture is undergoing some kind of degradation, but when you look at the alternatives—the truncation of the pan-and-scan technique, the bulky black or grey framing of letterboxing—anamorphic looks like the best option. It more or less fills the screen and the picture looks as sharp as the delivery medium will allow.

Actually, "squeezing" isn't the best way to describe the anamorphic process. Technically, what's happening is that an anamorphic DVD devotes more scan lines to the image and fewer to the blank bars framing the screen. Then the video display, correctly set to display an anamorphic picture, and playing anamorphic software, will reproduce a full-screen picture with greater resolution. In the process, it also corrects the misshapen image that would appear without anamorphic decoding.

Many sets recognize the anamorphic mode and adjust to it automatically. However, beware of sets that automatically lock into 16:9 mode when receiving 480p progressive DVD input. They may incorrectly apply the anamorphic process to non-anamorphic material (for example, an older movie such as *Casablanca*).

The widescreen dilemma

Is widescreen video an inevitable form of progress, a major step forward? For the most part, yes, but it comes at a cost.

Entering the widescreen world, by buying a widescreen set, involves a tradeoff. Yes, you'll be able to watch 16:9 anamorphic DVDs and widescreen Blu-rays in all their glory. But 4:3 material, including many favorite older movies and current TV shows, doesn't occupy the full width of a 16:9 set. You'll have to accept either blank bars at left and right or (the greater of two evils, in my opinion) horizontal stretching of the image. Disconcertingly stretched images are all too common in stores—retailers really should know better.

Simply put, nonwidescreen material looks better on a nonwidescreen set, while widescreen material looks better on a widescreen set. As a practical matter, if your TV museum includes a 4:3 set, you'll want to learn how to set your DVD player or set-top box in the **4:3 letterbox** mode, which will correctly fit an anamorphic 16:9 image to the 4:3 screen by eliminating a few scanning lines. If you own a 16:9 set, Blu-ray is the best option, anamorphic DVD is second best, followed by the old-style letterboxed version, though it won't have as much resolution.

Panned-and-scanned or full-screen video transfers—especially on low-resolution VHS—are lowest in the videophile pecking order.

Your DTV may come with several aspect-ratio modes to tailor various programming shapes to its screen shape. It may place blank bands at the top/bottom or left/right. It may blow up the image to fill the screen, cropping out some material. It may even do the latter in combination with some deliberate bending of shapes at the sides. You can choose the mode that suits the program material and your own tastes.

Is 21:9 the future of aspect ratio?

HDTV is getting even wider with a **21:9 aspect ratio**. Vizio and LG have introduced 21:9 sets. These sets are well equipped to display movies in ultra-wide aspect ratios such as classic CinemaScope at 2.35:1.

Is this progress? When it comes to width, is more better? It's important to remember that when a set gets wider relative to its height, its height gets shorter relative to its width. And it handles program aspect ratios with different kinds of masking. A 21:9 set will minimize or eliminate horizontal bars when used to watch ultra-widescreen movies. However, with 16:9 movies and TV shows, it will introduce vertical bars where previously there were none. And with older 4:3 programming, those vertical bars will be enormous. Using the stretch mode simply makes matters worse by distorting the shapes of onscreen objects. Do you really want a car wheel to look like an oval?

Higher horizontal resolution is one advantage of a 21:9 panel. A Vizio model, for example, has 2560 by 1080 pixels (versus the 1920 by 1080 of the HDTV standard). But when displaying a 2.35:1 movie in the HDTV standard, the signal's image area is 1920 by 817, with the rest used for top and bottom bars. So the signal has lower vertical resolution. This gets upconverted to the set's 2560 by 1080 pixels.

For TVs, file 21:9 under "for widescreen movie lovers only"—and not even for all of them. However, projector users may wish to explore variable-width screens with adjustable masking. Some even prefer to use different screens for different aspect ratios.

Curved screens

A few LCD and OLED TVs—from Samsung, LG, and Sony, so far—have curved screens. In theory, this is supposed to make the TV more immersive and panoramic, with improved depth of field. If you sit close up, the wraparound activates your peripheral vision. Wider viewing angle is said to be another advantage. With any TV, your eyes perceive the screen differently as you move out of the sweet spot, with the rectangle taking on a keystone shape. A curved screen changes this perception (whether you like it better may be a separate question). Curved screens are found only on large models, where the curve makes a difference, and only on pricey models, because of added manufacturing cost.

UHDTV and 3DTV

As flat-panel TV prices have plummeted, TV makers have desperately sought new ways to sell high-end sets. The latest technology on which they're pinning their hopes is UHDTV. Their previous attempt to rekindle interest in new TV purchases was 3DTV, though its cachet has waned, and today it's considered more a feature than a product category.

Ultra High-Definition Television: a definition

The latest high-end video buzz is what the Consumer Technology Association (formerly the Consumer Electronics Association) calls **Ultra High-Definition, Ultra HDTV,** or **UHDTV**. The informal (but widely used) nicknames are **4K x 2K** or **4K** for short. Note that this book sticks with the industry-approved Ultra HD and UHD monikers.

In 2012, the CTA (formerly CEA) initially defined UHD displays as having at least eight million pixels, 2160 by 3840, with aspect ratio of at least 16:9, and at least one UHD-capable non-upconverting video input.

In 2014 it added more requirements: UHDTVs must upconvert HD video to UHD resolution. They must have one or more HDMI inputs displaying 2160 by 3840 video progressively at 24, 30, and 60 frames per second. Color must adhere to the ITU-R BT.709 **color space** and may support more recent and wider color standards. **Color bit depth** must be at least eight bits. Note that some of these standards are pretty minimal. The BT.709 color space dates from 1990 and there are much higher standards for color bit depth. However, CEA's intention was to implement UHD using the existing technologies available to manufacturers in 2014. The quality of those technologies rises steadily.

Then there are additional demands for internet-connected UHD displays: They must meet all the requirements above, decode the **HEVC H.265 (High Efficiency Video Coding)** compression format, handle multichannel audio, receive IP-delivered video through wi-fi, ethernet, and other connections, and support IP video via services and apps.

HDR: an improvement over early Ultra HD

There is more to UHDTV than resolution and additional pieces still need to fall into place. Not all sets currently marketed have these pieces.

Among the most crucial improvements is **HDR (high dynamic range)** video, an add-on that improves the brightness and black level of an Ultra HD picture. Since black level is a traditional weakness of the LCD sets most people buy, this is a big deal.

HDR video at its best intensifies bright objects, like a sunlit sky, while improving shadow detail, preventing objects from being camouflaged and lost in areas of low brightness. It also provides more gradations between the extremes of bright and dark. In theory HDR systems now being devised, taken to their ultimate extensions, would produce an image bright enough to be painful to the human eye. But think of that as headroom; the whole capability won't be used. A system exceeding the limits of our vision is better than one that doesn't measure up to our vision.

HDR formats are **metadata**-based. That means the HDR data stream travels separately from the main picture stream. When an HDR-compatible set receives both, it adds the HDR stream to the main stream to produce an enhanced picture. There are several metadata HDR formats, some of which add color-related improvements (see next section). Some HDR formats, including Dolby Vision, are proprietary. Others are open standards such as SMPTE ST 2086, MaxFALL, and MaxCLL. HDR requires the **HDMI 2.0a** interface, which has been updated from 2.0 to pass the metadata.

Most major manufacturers offer HDR-enriched sets and others are expected to follow.

HDR involves improving the software as well as the hardware. Blu-ray titles incorporating HDR standards will start with three Warner Bros. titles encoded in Dolby Vision. The first HDR capable cable box came from Comcast in mid-2016, though oddly enough, it wasn't UHD capable. The cable industry will have to get its act together if it doesn't want to lose quality-conscious viewers to streaming.

The studios will incorporate HDR into their remastering of back catalogue. Samsung is offering consumers a UHD Video pack including two full-length movies encoded in HDR. Netflix is shooting original programming in HDR, presumably to make it available in its UHD streaming. The HDR revolution has just begun.

The HDR format war: HDR10 vs. Dolby Vision

HDR picture-improvement technology for Ultra HD is taking two forms recognizable to the consumer: HDR10 and Dolby Vision. Both use metadata embedded in content to tell the video display how to tweak itself for best results. But they do this in different ways.

HDR10 is the dominant one, at least for now. It is based on open-source SMPTE ST 2084 technology and has been adopted by Hisense/Sharp, LG, Philips, Samsung, Sony, and Vizio. Compatible streaming is available from Amazon, Netflix, and Sony's ULTRA 4K. If a TV isn't specified as having Dolby Vision, it probably has HDR10. One advantage of HDR10 is that it can be added via software update, whereas Dolby Vision requires a dedicated chip to built into the TV.

Dolby Vision, the HDR technology from Dolby Labs—better known for its surround sound formats—has been in the making since 2007 when Dolby bought BrightSide Technologies. DV made its debut in 2014. In lieu of the SMPTE technology, it uses a **perceptual quantizer** that dynamically maps content according to the specific brightness and color capabilities of the TV, using a chip in the TV to send that information to the source component. Dolby Vision has been adopted by LG, Philips, TCL, and Vizio. Dolby Vision content is available via the Netflix and Vudu smart TV apps and Amazon will follow; studios supporting the format include MGM, Sony, Universal, and Warner. Dolby Vision is optional in the UHD Blu-ray spec and no compatible BD players were available at this was being written in mid-2016.

With both formats used by hardware makers LG, Philips, and Vizio, as well as content providers Amazon and Netflix, this may be the kind of format war that ends in a ceasefire followed by mutual coexistence.

Color-related Ultra HD improvements

HDR is not the only way in which UHD pictures are about to get better. There are several other improvements relating to color. Some are being packaged into HDR formats, while others may be TV features

31

standing on their own.

Color gamut, or **color space**, is the range of colors. It is depicted in color charts as a triangle whose corners reach toward red, green, or blue. A subset of the triangle, often pictured as a triangle within the triangle, shows how much color can be supported by certain standards or equipment. In older color gamut standards, such as Rec.709, the subset is relatively small, indicating limited color. In newer ones, the subset expands, improving certain colors, especially red and purple. The incoming standards to replace Rec.709 (also known as BT.709) include DCI-P3, which is already prevalent in digital cinema projection, and the more ambitious Rec.2020.

Color bit depth is the range of shades within colors. The existing 8-bit color depth embraces up to 256 different shades for red, green, and blue with each specific color having a number. More colors would be better, of course, which is why CEA's HDR standard is moving to a 10-bit color depth, which would offer 1024 shades per color, as included in the HEVC compression standard. There might even be a next step to 12-bit color depth, as supported by Dolby Vision. The benefit of greater color bit depth is less color banding, reducing abrupt transitions from one shade to another.

Color compression or **color subsampling** is a necessary evil of digital video that saves bandwidth by reducing the level of chroma (color) information relative to luma (brightness) information. The less compression, the better. CEA requires only 4:2:0 compression, which halves both horizontal and vertical chroma resolution, but TV makers are moving to 4:2:2, which halves only horizontal chroma resolution, leaving vertical resolution uncompressed. The next step would be to 4:4:4, which means no compression at all.

The HDMI 2.0 and 2.0a interfaces, with their increased bandwidth, are required to pass these signals.

UHD on Blu-ray

Hardware isn't the only thing UHD needs to be viable. It also requires UHD production, consumer formats, a standardized interface, and software releases. Most big-budget movies are produced in UHD but most TV shows are not. To get those movies into your UHD set, the best bet is UHD Blu-ray, which arrived in late 2015. For more information see "Picture & Sound Sources/Disc players/Ultra HD Blu-ray."

Other UHD programming

UHD is available via video streaming from Netflix, Amazon, YouTube, and Vimeo, among others. Satellite operators DirecTV and the Dish Network have announced UHD services. So has Comcast, the nation's largest cable operator. While all this gets up and running, TV makers are filling the gap by providing hard-drive-based devices to UHDTV buyers. Sony and Samsung are among them.

The HDMI 2.0 (and up) interface improves UHD frame rate from 30 to 60 per second—look for it in UHDTVs and associated gear.

Consumer alert: HDCP 2.2 anti-copying technology

HDCP 2.2 may have an impact on what UHD programming you're allowed to watch on your UHDTV—and the news here is not all good.

The problem is not HDCP, or **High-bandwidth Digital Content Protection**, per se. Older generations of this anti-copying technology routinely do the copyright handshake for HDTVs, Blu-ray players, and other HD gear. The problem is that the latest version 2.2, with its tougher encryption keys, is being implemented unevenly in first-generation UHDTVs. Some have it, some can upgrade their firmware for it, but some will never be able to accommodate it. *You cannot assume that any UHDTV or HDMI 2.0 device has HDCP 2.2.*

So look for HDCP 2.2 on the spec sheet and don't buy any set lacking it. Also, if you plan to run UHD signals through a surround receiver or soundbar, they will need to pass it too—and as this edition is being written, only some of them do. But you needn't worry about HDCP 2.2 preventing your existing Blu-ray player and other source components from feeding a UHDTV. Nor should you worry if you have a non-UHD HDTV. HDCP 2.2 affects only UHD sources feeding a UHDTV.

ATSC 3.0: UHD over the air

The Advanced Television Systems Committee is revising the HDTV broadcast standard to add UHD in **ATSC 3.0**. Incidentally, there was never an ATSC 2.0. The ATSC folks decided to move straight from 1.0 to 3.0, kind of the same way Windows went from 8.1 to 10. It's bigger and better than ever—the number says so!

Unlike ATSC 1.0—the original DTV transition—ATSC 3.0 won't require an act of Congress, just the approval of the FCC. The DTV transition started with broadcast TV and progressed into other media, including streaming; the UHDTV transition started with streaming and is progressing into other media, including broadcast. The standard is still being formed and tested both in the U.S. and abroad.

The ATSC 3.0 specs will look a lot like the UHD specs, with the same 4K resolution, HEVC H.265 video compression, up to 120 Hz frame rate, some form of HDR, wide color gamut, and possibly some form of object-based surround sound similar (but not identical) to Dolby Atmos. It might even support 3D. And it will be designed not just for broadcast but for internet transmission as well.

ATSC 3.0 will not be backward compatible with 1.0. That doesn't mean your existing DTV tuner will become obsolete overnight. But as 3.0 broadcasts begin, you'll need to add a 3.0 compatible tuner. It might be a set top box, a dongle, or—the least awkward option—incorporated into a new TV. In that respect, the ATSC 3.0 transition will resemble the original DTV transition. This time, however, broadcasters and the electronics industry are both urging the FCC to make compliance voluntary.

This new broadcast standard comes just as the FCC is auctioning off part of the DTV spectrum to make way for mobile broadband. Because ATSC 3.0 is not backward compatible with 1.0, TV stations will have to be creative in accommodating both at once, perhaps hosting one another's signals. But they'd also be able to deliver more channels, interactivity, and datacasting, perhaps bringing in more revenue. For those who live in the path of tornados, an Advanced Warning and Response Network would deliver geo-targeted alerts and evacuation maps. Who knows, maybe ATSC 3.0 will save a life or two.

Ultra HD Premium Certification

A group of TV makers, movie studios, and other content providers formed the UHD Alliance to create an **Ultra HD Premium** certification for TVs and software. Members include LG, Panasonic, Samsung, Sony, Disney, Fox, Universal, Warner, Amazon, DirecTV, Dolby Labs, Microsoft, and Netflix.

Their requirements include 2160 by 3840 pixels, 10-bit color depth, at least 90 percent of the P3 color gamut (less extended than the superior Rec. 2020), and HDR with SMPTE ST 2084. LCD sets with local dimming would be required to support a minimum brightness of 1000 nits with black level of no more than .05 nits (a 20,000:1 contrast ratio). OLED sets would be required to have minimum brightness of 540 nits with black level of no more than .0005 nits (a 1,080,000:1 contrast ratio).

Do you need UHDTV?

UHDTVs now dominate the top end of major TV lines. But with 1080p sets already routinely providing superb pictures, do we need this? Don't scoff. UHD is already becoming the standard in digital cinema projection (with resolution of 2160 by 4096), which is fast taking over from film projection. Naturally makers of high-end home projectors are making UHD available to bleeding-edge videophiles. Sony and LG introduced the first UHD flat-panel LED-LCD TVs, both 84 inches, and others are following. Let the land rush begin.

For a typical home HDTV, UHD may be overkill. But on a big enough screen, at a close enough seating distance, and if your eyesight is good enough, it might make a difference. Also, movies shot in UHD look better even when downconverted to 1080p. For those viewing 3DTV with passive glasses, a UHD home projector can provide full 1080p resolution (otherwise only available with active glasses). UHD may also provide advantages other than resolution—in color or quality of compression.

So yes, UHD is a big deal, even though not everyone needs it. And be warned that UHD sets don't all perform alike. A good HDTV is better than a lousy UHDTV. Read reviews for informed assessments.

After 4K: is 8K next?

Even as TV makers are assimilating 4K Ultra HD, **8K** technology is rearing its head. NHK, the Japanese broadcast authority, ran an experimental broadcast with the 2014 World Cup, beaming 8K signals to Japan and Brazil. In 2016 NHK broadcast the Rio Olympics in 8K. Further satellite broadcasts are set for 2016, 2018, and 2020, when Tokyo will host the Olympics. By then NHK will be using broadcast technology developed in partnership with Sony and Panasonic.

Do you need 8K? The short answer is no. 4K UHD already exceeds the limits of the human eye with room to spare. Do TV makers need this? They always need something new, but as with the 3D fiasco, a new technology doesn't always take root. Can TV makers even manufacture credible 8K? Well, for the moment, they're still struggling with getting all the pieces of 4K into place, and many 4K sets are mediocre performers despite their resolution specs. You still get what you pay for.

35

There may be a place for 8K in production, archiving, and theatrical distribution. But unless something changes dramatically, home users needn't worry about planning an upgrade path to 8K. Even 4K UHD is already overkill for smaller sets.

So much for 4K and 8K TV. Now we turn our attention to the last flavor of the month, 3DTV. Spoiler alert: It's already on its way out.

History of 3D in photography and movies

A 3D image, in general, creates an illusion of greater depth perception by presenting each eye with slightly different images. This is known as **stereoscopy**. It goes back to the 19th century, when it was first used for still images. 3D for motion pictures made its debut in 1922 with *The Power of Love*. A major wave of 3D films arrived in the 1950s including *Bwana Devil* (1952), *House of Wax* (1953), and Alfred Hitchcock's *Dial M for Murder* (1954). 3D has continued to arrive in waves, with another major one in the 1990s. Of course the current cinematic and video revival of 3D has been led by *Avatar* (2010).

Basic types of 3DTV

What kind of TV is the best match for 3D processes? The leading contender was plasma, due to its inherently high motion performance. LED-backlit LCD sets can also produce solid 3D effects. Another alternative is DLP. For 3D front-projection, use a higher-gain screen to offset the reduction in brightness.

Because they require high-end video processing, 3DTVs also work well in 2D mode. Some provide **2D-to-3D conversion**, though this does not provide the optimum 3D experience.

There are many types of 3D technology. One way to sort them out is by looking at the glasses required to decode images. 3D glasses may be **active** or **passive**. The most common type of 3D in its current video revival is **frame sequential**, which displays alternating frames for a viewer's left and right eyes. Battery-powered **active-shutter glasses** are synchronized by an infrared or RF transmitter on the 3DTV and a receiver in the glasses. On the passive-glasses side, other types of 3D include **anaglyphic**, which uses glasses with different-colored lenses, red and cyan; **polarization**, which uses passive polarized glasses; and **autostereoscopy**, which requires no glasses.

If you prefer 3D with passive glasses—which eliminate batteries

36

and syncing problems—check out **3D Ultra HD** models from LG, Sony, and Toshiba. They use the extra resolution available in UHD to counter the loss of resolution inherent in passive 3D.

3DTV technology is baked into the panel, so whatever TV you buy will dictate the kind of 3D available to you. For instance, Panasonic and Sony use frame sequential technology with active-shutter glasses (as did Samsung before exiting the 3D category)—and each has its own proprietary version. Vizio, LG, and Toshiba use polarization with passive glasses that look like sunglasses. Active 3DTV can deliver full 1080p resolution in both 3D and 2D modes though the glasses can be expensive. Passive 3DTV has the advantage in cost but requires a patterned barrier to be attached to the screen, which pollutes 2D performance.

3DTV formats via HDMI

The HDMI interface is the preferred method for getting 3D images from 3D source components to 3DTVs. The first version of HDMI to handle 3D was 1.4. For 3D in Ultra HD you need HDMI 2.0 and up.

HDMI 1.4 supports a variety of 3D formats including frame packing, checkerboard, interlaced, line-by-line, alternate frame, and 2D+depth. For 3D video delivered to the home, the two most significant frame-packing formats are **top-and-bottom** and **side-by-side**. Visualize a frame split in half, either horizontally or vertically, and you'll get the idea. The 3DTV processors separate and scale this signal into individual left and right images. The active-shutter glasses direct each image to the appropriate eye, resulting in a single 3D image.

Blu-ray, with its capacious bit bucket, uses **frame packing**, the best possible form of the top-and-bottom method, with full 1080p resolution (1080 by 1920 pixels) in each eye.

DTV channels using top-and-bottom 3D cut vertical resolution in half. So 1080i video, normally 1080 by 1920 pixels, goes from 1080 vertical pixels to 540; 720p video, normally 720 by 1280 pixels, goes from 720 vertical pixels to 360, a serious loss of resolution.

Bandwidth-starved media like satellite and cable use side-by-side 3D and cut horizontal resolution in half. So 1080i video, normally 1080 by 1920 pixels, goes from 1920 horizontal pixels to 960. At present there is not a 720p side-by-side format in HDMI 1.4a.

HDMI 1.4a adds a few more 3D formats including frame-packing top-and-bottom at 2205 by 1920 pixels, which supports full 1080p resolution. For video games there's a full 720p version. To ensure maximum

3D compatibility, make sure any TV, Blu-ray player, or source component you buy meets the HDMI 1.4a specs.

The lemmings approach the cliff

I wish I could enthusiastically recommend 3DTV to my readers but I have several reasons not to. One is a relative lack of industry-wide standards. Each 3DTV maker seems to have its own approach, so you can't be sure one's glasses will work with another's 3DTV.

Another problem with 3D, at least for a minority of viewers, is that it simply doesn't work. Not everyone's eyes and brain mesh with the various 3D technologies.

There are health concerns. A 3D warning given on Samsung's Australian site blurted out a whole laundry list of complications including epileptic seizure, stroke, altered vision, lightheadedness, dizziness, eye or muscle twitching, confusion, nausea, loss of awareness, convulsions, cramps, disorientation, motion sickness, perceptual aftereffects, eyestrain, decreased postural stability, damage to eyesight, headaches, and fatigue. While these warnings may be overblown, parents should limit the number of hours kids are exposed to 3D. If watching 3D bothers you, stop watching it. People with serious eye- or brain-related health problems may wish to avoid it altogether.

But the most persuasive argument against 3D is aesthetic. It came from Pulitzer-winning film critic Roger Ebert, who wrote: "When you see Lawrence of Arabia growing from a speck as he rides toward you across the desert, are you thinking, 'Look how slowly he grows against the horizon' or 'I wish this were 3D'?" Perhaps 3D's effect is not so much depth enhancement as depth exaggeration. Time-honored cinematic techniques, such as the use of perspective, may have the artistic high ground.

Home theater has been greatly improved by technologies introduced over the past decade or two—including HDTV, flat panels, and lossless surround—but 3DTV is not one of them. It is the most overrated new (actually not so new) video technology of recent years. And its implementation has been scattershot at best, with dueling screen technologies and numerous signal formats.

3D shows signs of slowing down. ESPN has killed its 3D sports channel. Expansion of 3D theaters has slowed. And the trickle of 3D movie releases slowed from a high of 91 in 2011, the year after *Avatar*, to 48 in 2014. TV makers seem to have shifted their focus from 3D to

Ultra HD. Samsung discontinued 3D for its 2016 line. Without a constant stream of movie releases and active TV lines, 3D is in danger of becoming irrelevant in home theater.

I fully understand that some readers do not share my skepticism. But unless you've fallen in love with 3D at the cineplex, feel free to give the home version a miss.

Smart TV, tuners, & cable

How you feed your TV determines not only what you watch but how good it will look and sound. This chapter will discuss program delivery features that can be built into a TV including internet, broadcast, and cable. Add-on components including satellite receivers and antennas are discussed in the "Picture & Sound Sources" chapter.

Smart TV

Smart TV is the industry's catch-all term for TV with internet (**IP-enabled**) features. Most prominent among these features is video streaming. There are subscription services like Netflix, pay-per-view services like Vudu, and free services like YouTube or TV network and station websites. Smart TVs may also offer **audio streaming** (Spotify, Pandora, vTuner) or **social networking** (Facebook, Twitter). While all these things can also be done by computer, smart TV tends to adopt a simpler interface that is more icon- and app-based, more like a tablet than a computer. This enables it to be operated with a remote control or phone/tablet app.

A smart TV isn't the only way to get smart TV features. Other internet-enriched a/v products include Blu-ray players, set top boxes, game consoles, and receivers. You don't need smart features in all these things but it is convenient to have them in whatever components you use the most. For example, if you like to watch movies on disc, but also like to stream, you might appreciate having video streaming in your Blu-ray player. If you're a gamer who watches the occasional movie, you might prefer to have video streaming in your console. If you're into music, it can be handy to stream internet radio from a surround receiver.

Smart TVs connect to a home network in a variety of ways. Wired

39

connections use ethernet or a powerline ethernet adapter. Wireless alternatives include wi-fi, Apple's AirPlay, Miracast for Android devices, Intel's WiDi, WHDI (Wireless Home Digital Interface), and others that are either failed or formative. Once the smart TV connects to the network, DLNA certification enables it to pull media from a router-connected PC. Adoption of HDMI version 1.4 has accelerated the spread of network features in DTVs by letting devices share an internet connection and exchange data in a standardized (as opposed to proprietary) manner.

Smart TVs can connect to certain devices without a network. Some smartphones and tablets use a wired non-network connection, the MHL variant of HDMI, to stream video and audio. Bluetooth offers a wireless device-to-device connection for audio streaming.

As consumers have "cut the cord" and moved away from cable and other traditional means of video delivery, network features are shifting from optional to indispensable. However, be warned that IP video is not necessarily high-def, and even nominally HD streaming can be low in quality. Blu-ray and antenna TV remain the highest-quality sources of HDTV, followed by cable, satellite, and IPTV.

DTV broadcasting

Friends, the world has changed. The first decade of the 21st century saw a major transition—the transition from analog to digital broadcasting. Analog broadcasting ceased on June 12, 2009 by an act of Congress.

Owners of analog TVs can still use their old sets with set-top boxes. As for cable, the Federal Communications Commission passed "must carry" rules mandating that cable systems carry all digital over-the-air channels and make them available to all sets, digital or analog—either converting the signal at the head end or providing convertors to subscribers. Brawls between broadcasters (who want to be paid for their signals) and cable operators (who want to pay as little as possible) are testing the FCC's determination to enforce these rules.

Digital television signals reach every major market in the U.S. To check the availability of DTV broadcasts in your area, consult the dtv.gov or tvfool.com websites.

To receive over-the-air signals, a video display needs a tuner. (Satellite, cable, and telco delivery require set-top boxes.) A **digital tuner** (or **ATSC tuner**) is one that receives digital channels. The **analog tuner** (**NTSC tuner**) is obsolete. Formerly, sets with digital tuners were called **integrated DTVs** while those without digital tuners were designated

DTV ready. Now any set called an **HDTV**, an **EDTV**, or an **SDTV** is presumed to be an integrated set with digital tuner built into it. Any set that can display DTV but requires a separate tuner is classified as a **monitor**. An outboard channel selector is still called a **tuner**.

Except in front-projection, DTV tuners have been legally required since 2006 by the FCC. Critics point to the fact that most Americans don't need over-the-air reception and say adding the cost of a digital tuner to every DTV is unfair. Others point out that even people who depend on satellite or cable service still use antenna reception in the bedroom or other parts of the home. An increasing number of post-cable cord-cutters use antennas exclusively.

Cable tuners

In addition to broadcast-TV tuners, DTVs also have cable-TV tuners (as did their analog forebears). Cable uses a **QAM** (quadrature amplitude modulation) **tuner**. Until recently a cable subscriber could receive basic cable (broadcast and other low-tier) channels using the set's QAM tuner with no need for a cable box or card.

However, the FCC—after decades of cable lobbying—has ended its **encryption ban**, starting with broadcast channels in 2012 and adding basic cable in 2014. Now cable ops can encrypt these channels, making the set's QAM tuner useless, and forcing use of a box. If you suddenly find your basic cable channels encrypted, you might be entitled to a box free for up to five years. The FCC also requires it for viewers on Medicaid. You must act within four months of the start of encryption.

There is one loophole to QAM encryption: Cable operators must make basic channels accessible via internet video to allow IP-based QAM set top boxes to keep working. This has been called the Boxee exemption due to that company's successful lobbying. Otherwise, if you want to watch HDTV without a box, your only option is to cut the cord and get an antenna.

Tuner features

Picture-in-picture (**PIP**) allows simultaneous viewing of two or more pictures on one screen. You can monitor one channel while watching a picture inset of another channel or a rented movie. Most sets offering PIP display the extra picture using a separate tuner or another source such as a disc or tape. **Dual-tuner PIP** packs both tuners into

41

one TV. Widescreen sets may have **picture-outside-picture** (**POP**).

Federal law requires that all TVs 13 inches and up have a **closed-caption decoder**. It displays printed material onscreen taken from the broadcast signal or from any captioned disc or tape. You may find additional captioned channels for foreign language, statistical displays, secondary analysis during news events, emergency announcements, weather, stock market reports, or program listings.

Closed captions are part of **Extended Data Services** (**EDS**). These can include elapsed time and name of program as you switch channels, set TV or DVR clocks, delay or extend recording when program times change, or display channel call letters and network affiliations.

Another federally mandated feature is the **V-chip**, which provides **parental control** and **channel lockout** functions. Concerned parents may block channels showing rated programs deemed unsavory for children. Should those same parents decide to unblock a channel at the midnight hour, they can restore a previously blocked channel with a numeric passcode.

Channel labeling liberates you from your cable system's meaningless channel numbering scheme, which may not even correspond to broadcast channel numbers. You can rename or renumber any channel to something that makes more sense to you, such as the station's call letters or network affiliation.

Auto channel programming scans through all channels to separate active from inactive ones. When surfers later step through channels, the tuner skips those with no signal, as well as any that have been blocked. It's not a bad idea to make a DTV set rescan channels on a regular basis to account for modifications in the broadcast signal. Station engineers are always tweaking something.

Stereo and surround via broadcast or cable

The tuner's role goes beyond channel selection and picture. It also delivers audio, potentially including surround or stereo. A DTV tuner delivers audio encoded in **Dolby Digital** (see "Surround Sound/Understanding surround standards"). If you want to receive analog stereo via cable—a must if you also want Dolby Surround sound— look for **MTS**, which stands for **multichannel television sound**.

Surround and stereo subtly enhance everything from talk shows to sporting events. When the stereo soundtrack carries Dolby Digital, shows like your favorite primetime fare, *The Late Show, The Tonight Show,* and *Saturday Night Live* take on a real you-are-there feeling, as though you were right there in the audience, in the heart of the crowd. Musical segments can sound almost as good as a compact disc.

SAP stands for **second audio program** and is found in all digital tuners and in obsolete MTS-equipped analog tuners. While not all broadcast stations or cable systems pass on these signals, some TV stations are becoming creative in their use of SAP. Some go beyond its obvious use for bilingual soundtracks to provide extra commentary during sporting events, uncut audio portions of R-rated movies, various kinds of background information, and audio-only news or weather services.

Volume control

Volume correction is a helpful feature. TV ads are typically broadcast near the top of the allowable limit, unlike the programs themselves, which vary between loud and soft. A circuit that automatically reduces the volume of blaring ads may save your remote-operating hand a lot of work. Dolby Labs, THX, and Audyssey have revived this old (but good) idea and license various forms of it for a/v products under the names **Dolby Volume**, **THX Loudness Plus**, and **Audyssey Dynamic Volume/EQ**, and **Audyssey Dynamic Volume TV**. TV makers should seriously consider adding one of these processing modes to their sets.

Such heroic measures are becoming unnecessary thanks to the Commercial Advertisement Loudness Mitigation Act (or **CALM Act**), which became law in 2010. This act of Congress requires the Federal Communications Commission to implement ad-volume rules already suggested by the Advanced Television Systems Committee.

Digital cable readiness

There have been roughly four generations of digital cable readiness. First came the ill-fated CableCARD, which was promised to make TVs and DVRs independent of cable boxes, though all but 617,000 of the 53 million CableCARDs deployed ended up in set top boxes. Then came the short-lived Tru2Way, the FCC's stillborn AllVid, and the forthcoming DSTAC standard. Let's review their sad history.

Digital cable readiness *is the law!* You can thank the FCC for this. Finally acting on a 1992 federal law, the FCC began enforcing an **integration ban** in July 2007. What this means is that the security function once exclusively provided by cable-provided boxes can now be provided by a card-enabled cable-ready TV, by a card-enabled cable box rented from the cable operator, or by a store-bought box—giving consumers wide latitude in how to handle their cable-TV connections. Your cable operator would prefer that you rent a box but can no longer compel you to do so. The following backgrounder will explain how we got to this point and where we might go in the future.

The term **cable-ready** has long been a battleground for government, television manufacturers, and the cable industry. Consumers need a simple way to connect a cable TV feed to a DTV and receive digital programming without having to go through a cable-company-supplied set-top box. Such a cable TV **plug and play** standard now exists thanks to an agreement signed by cable operators and TV makers on December 12, 2002. Plug-and-play DTVs and other products are designated **digital cable ready** or **DCR**. The more technical term is **unidirectional digital cable**. The standard was supposed to have been enabled by July 1, 2004 in all cable systems with an activated channel capacity of 750MHz or greater though some cable operators dragged their feet.

The process is quite elegant in practice. A digital cable ready product can access scrambled cable channels, as well as unscrambled digital and analog channels (but not pay-per-view or video-on-demand). To receive premium and other scrambled premium channels, the user inserts a security device known as a **CableCARD** into a slot that could be built into a set-top box, DVR, or DTV. Most CableCARDs are deployed in boxes.

The unidirectional digital cable agreement provides for home networking and recording of digital video with some limitations on some content. One option that may be exercised by the cable operator on pay-per-view content is **copy never**. That may sound draconian but actually it does come with a loophole that would permit hard-drive-based digital video recorders to store paused programming for a minimum of 90 minutes. Another option is **copy one generation**, which allows material to be copied to a recording device (but not recopied). Up to 64 simultaneous first-generation copies are allowed—just don't try copying a copy.

In a temporary victory for consumers, one controversial feature that digital cable ready products initially weren't required to include is **selectable output control**, which would enable cable operators to use cable-ready gear to completely shut off recordable 1394 outputs and analog outputs including component video. Also prohibited—but only for unencrypted broadcast channels—is the ability to **down-res** (reduce the resolution of) HD signals delivered to the component inputs of millions of early HDTVs. The down-res of other channels is not addressed in the agreement.

Hollywood lobbied the FCC for a waiver that would let selectable output control be used for video-on-demand delivery of major movie titles. The FCC eventually gave into this demand in 2010, providing a limited waiver that would allow use of SOC for 90 days or until disc release. The waiver was extended in 2012.

The initial digital cable ready agreement is for unidirectional (one-way) service and has been approved by the FCC. A newer type of cable card, known as **Tru2Way**, provides for **bidirectional** products—those with a return path, allowing two-way communication between the cable system and the box or set. Bidirectional products support additional services such as video on demand, interactive program guides, video games, and electronic commerce. Tru2Way was not officially sanctioned by the FCC but was adopted by several DTV and DVR makers including LG, Panasonic, Samsung, Sony, and TiVo. The cable companies were more open to the Tru2Way standard because it allowed them to market new services. But the last TruWay supporters gave up in 2010.

Another approach to digital cable readiness never got past the theoretical stage. The FCC suggested an **AllVid** gateway to eliminate set top boxes. The cable operators opposed it and it went nowhere.

TV makers have offered digital cable ready products though their enthusiasm and support have waned over the years. They feel burned by the cable industry's halfhearted compliance (and frequent non-compliance) with the original CableCARD agreement. But, as discussed in the opening of this section, the FCC integration ban is keeping the concept of digital cable readiness alive.

Cable readiness: the next generation

Under orders from Congress, the FCC is now considering a next-generation replacement for the CableCARD. The first step was to empanel the **Downloadable Security Technology Advisory Committee (DSTAC)** in 2014. It considered both existing and new technologies from various parties. After DSTAC delivered its report in 2016, the FCC voted 3-2 along party lines to start a rulemaking process that would culminate in unlocking the dreaded cable box. It allows for either hardware- or software-based approaches. The new system would break down into three data streams: finding available programming, figuring out what you're authorized to view or record, and getting it into your TV.

Several approaches emerged. One front runner was **VidiPath**, from the Digital Living Network Alliance (DLNA), which has been adopted by three major cable operators. Another came from a coalition formed by Google, TiVo, Vizio, and the advocacy group Public Knowledge. Another came from the National Cable & Telecommunications Association, the cable industry's lobbying arm, which recommends an app-based system based on HTML5 open web standards. The FCC will take a couple of years to pick through options, and might deliver a verdict by 2018—unless Congress interferes, as it is wont to do.

It is hoped that the new standard will have broader support than the CableCARD. One big question is whether it will end up being deployed in rented boxes (as cable operators might prefer) or directly in sets (as consumers might prefer). The original CableCARD was designed for cable-ready TVs, but miraculously ended up deployed overwhelmingly in boxes, allowing the cable industry to rake in $20 billion per year by charging subscribers an average of $232 per year for rented boxes. We can do better than that. But will we? Cable readiness has long been the train that never quite arrives at the station.

DTV connections

Most readers of this book would be better off using a receiver for their switching needs. Therefore the big issue in TV connectivity is not the number of inputs, but the kind of video inputs available, which will influence the video performance of your home theater system.

- HDMI/DVI input
- HDBaseT interface
- MHL interface
- DisplayPort interface
- IEEE 1394, DTV Link, FireWire input
- component video input
- RGB, RGBHV input
- VGA input
- S-video input
- composite video input
- RF input

- CableCARD slot
- front-panel video input
- digital audio input (optical or coaxial)
- stereo analog audio inputs
- audio/video outputs
- RS-232
- 12-volt trigger
- network connection
- WiDi streaming

The first two forms of high-quality digital connection to arrive on the back panel were HDMI and IEEE 1394. HDMI has become the dominant one, eclipsing 1394. To allay Hollywood's concerns, and to spur the growth of HDTV programming and disc releases, both come with digital rights management. Therefore HDMI comes with a form of DRM called HDCP, and IEEE 1394 comes with DTCP.

HDMI (**High-Definition Multimedia Interface**) carries UHDTV, HDTV, and multichannel audio. One attractive thing about HDMI is that the plug and jack take up a lot less real estate on back panels than other HD-capable options. HDMI was developed by Silicon Image though the HDMI Founders Group includes many major TV makers, movie studios, satellite TV operators, and CableLabs, the cable

HDMI (left) is the universally accepted audio/video connection for UHD and HD displays and source components. The latest version is 2.0. MHL (right) is a variation of HDMI that connects Android smartphones to HDTVs.

47

industry's R&D arm.

HDMI is an evolving standard that comes in several versions. Search spec sheets to find out what you're buying. The original **HDMI 1.0**, released in 2002, supports video and stereo audio, not surround. **HDMI 1.1** adds Dolby TrueHD, Dolby Digital Plus, and DTS-HD High Resolution Audio (cutting-edge surround standards that are significant parts of the Blu-ray disc format) and DVD-Audio (a high-resolution music format). **HDMI 1.2** adds Super Audio CD, the main rival to DVD-Audio among high-res disc formats, and is more computer-friendly. **HDMI 1.2a** adds CEC (Consumer Electronic Control) which allows components in an a/v system to respond to commands in a coordinated way. Several manufacturers support CEC, often disguising it with phony proprietary names to take credit for something they didn't invent. **HDMI 1.3** came out in 2006 and supports higher bandwidth, greater color depth, a new mini-connector for camcorders, and automatic lip-sync to keep soundtracks coordinated with faces speaking on-screen. It also supports DTS-HD Master Audio, the dominant lossless surround format in Blu-ray. **HDMI 1.3a** makes small modifications in CEC, color depth, and other areas. **HDMI 1.3b** makes no functional changes for the consumer.

HDMI 1.4 brings many improvements. It adds an **HDMI Ethernet Channel**. That helps connected devices share an internet connection, and they can exchange data at 100 Mbps, enough for any IP-based application. An **Audio Return Channel** reduces cabling by letting an HDTV pass the audio stream back to a receiver for surround decoding. Also supported is **UHD** resolution. There's an expanded color space, support for **3D** media, a **Micro HDMI connector** that's half the size of the **HDMI Mini connector**, and an **Automotive Connection System**. HDMI 1.4 specifies three kinds of HDMI cable. See the "Cables" chapter or HDMI.org for details. **HDMI 1.4a** mandates a fuller set of 3D formats for movies, TV broadcasts, and games. HDMI also now supports **latching connectors** which prevent the plug from falling out of the jack. These may use either a magnet or a mechanical prong. Some HDMI devices support **InstaPrevue**, which shows the content of HDMI sources through picture-in-picture insets. This makes switching easier by substituting motion content for dry input names or numbers. HDMI 1.4b adds support for 3D 1080p at 120 Hz, allowing frame-packing 3D at 60 frames per second.

HDMI 2.0, adopted in 2013, is not the first version to support UHD video, but it raises the frame rate from 30 to 60 frames per second, a considerable improvement that allows for 3D UHD as well as higher 2D quality. It also doubles bandwidth to 18 gigabits per second, supports the ultra-widescreen 21:9 aspect ratio, can deliver dual 1080p video streams to one screen, dynamically synchronizes video and audio streams to eliminate lip-sync delay, and supports up to 32 discrete audio channels for next-generation surround systems, with audio sampling rates up to 1536 kilohertz, enough for true high-resolution audio. The spec is backward compatible. It requires a high-speed (as currently defined) HDMI cable but not a new type of HDMI cable. **HDMI 2.0a**, the latest version, passes the metadata needed for HDR video.

The more functional versions of HDMI are now entrenched. If all you want is top-quality HD video and audio, here's what you need to know about HDMI right now—get version 1.3 or higher. If you want 3D, make that 1.4a. If you're into UHD, go for 2.0. For better UHD with HDR—strongly recommended—make it 2.0a.

Be warned that HDMI 2.0 does not automatically include **HDCP 2.2**, the anti-copying technology being used for Ultra HDTVs and UHD source components. Make sure any UHDTV you buy, as well as any other devices in the signal chain, has HDCP 2.2.

HDMI brings a couple of problems. Though not uncompressed, the signal is fat, with a lot of extraneous data added to make it deliberately unwieldy. As a result, it can't travel far, and the cables can be expensive. The **HDBaseT 1.0** standard addresses those problems by translating HDMI to a format that can be carried by standard (and dirt-cheap) Cat5e or Cat6 cables with RJ-45 connectors. HDBaseT supports video up to UHD for distances up to 328 feet. Another major advantage is that the standard supports power and internet as well as video and audio connections. **HDBaseT 2.0** supports plug-and-play switchers for network-connected HDMI sources such as video components, game consoles, security cameras, and wi-fi devices, making them controllable via tablet or smartphone. HDBaseT developers include LG and Samsung, and it appears in some of their TVs, plus Integra surround receivers and several brands of projector.

HDMI is the go-to interface for audio/video gear and computer monitors but its connector is a bit large for smartphones—and that can be a problem if you want to connect your phone to your home theater system. That's why the MHL Consortium has developed the **MHL (Mobile High-Definition Link)** interface. It does not have a standardized plug or jack though early implementations have used 5- or 11-pin mini-USB plugs at one end and HDMI at the other. **MHL 1.0** supports uncompressed 1080p video, 7.1 channels of uncompressed audio, and HDCP copying restrictions (see below)—so connected devices shouldn't have any problem doing the ol' copyright handshake. It also passes power from the display to the mobile device, so if you're watching a movie using a smartphone as the source, the latter won't poop out halfway through the show (though be warned that some implementations skip the power connection). **MHL 3.0** supports UHD, a high-speed data channel, improved remote protocol, higher power charging, HDCP 2.2 copy prevention, Dolby TrueHD and DTS-HD surround, and support for multiple displays, touchscreen, keyboard, and mouse. MHL is supported by several brands of TV, surround receiver, and smartphone. The new **superMHL** supports up to 8K video at 120 Hz, HDR, up to 48-bit color depth, wider color gamut, Dolby Atmos and DTS:X surround, the ability to link multiple devices, power charging up to 40 watts, content from a single device on multiple displays, a new reversible connector, and support for superMHL (six streams), HDMI Type-C (four streams), HDMI Type-A, micro-USB, and proprietary connectors. It is backward compatible with earlier versions. To use MHL with non-MHL devices, you'll need an adapter or dock. For a list of compatible devices, check out mhltech.org/devices.aspx.

The forerunner of HDMI is **DVI-HDCP (Digital Video Interface with High-bandwidth Digital Content Protection)**, a variation of DVI-D plus copyright protection. You may see DVI's larger (much larger!) plug or jack on older video and computer gear. HDCP, the content-security scheme for DVI, also serves the same function in HDMI. DVI and HDMI can be physically bridged by an adapter—but the two connected devices will work together only if they share the same signal and protection protocols. That can be an iffy proposition.

DisplayPort is a potential alternative to HDMI as an a/v interface though currently it is mainly for computer monitors. It is not compatible with HDMI. Version 1.4 supports 4K (UHD) up to the 120 Hz frame rate and 8K up to 60 Hz, with "virtually lossless" video compression up to 3:1, and 32 audio channels with a sampling rate of 1536 kHz. It does not yet have traction in mainstream a/v gear.

Another digital interface known as **IEEE 1394** is now rarely used in a/v gear. IEEE 1394 was named for the standard-setting body (and decree) that originated it, though it also has such proprietary names such as Apple's **FireWire** and Sony's **iLink**. The Consumer Technology Association and the National Cable Television Association have agreed on the term **DTV Link** as well as the related terms **DVD Link, Web Link,** and **D Link** (the last one is for camcorders). In one form, 1394 supports networking and recording of video signals—making it far more consumer-friendly in those respects than HDMI or DVI. It's also been used by a few manufacturers to connect universal disc players to surround receivers. In both spheres, implementation of IEEE 1394 has been stunted by the entertainment industry, which is militantly opposed to the possibility that digital copying may erode revenues. For that reason, an early-generation DTV set-top tuner (the Panasonic TU-DS50) is in great demand because it's one of the few that can feed the IEEE 1394 inputs of a high-def video recorder without the hindrance of anti-copy flags. Unfortunately, that model has gone out of production, and the unfettered IEEE 1394 interface is available only in PC tuner cards. Most products with 1394 interfaces support **DTCP** (**Digital Transmission Copyright Protection**). 1394-DTCP has at least the nominal support of most major TV makers, the Consumer Technology Association, plus the Warner and Sony studios.

Component video is the highest-quality analog video connection and the only one that's HD-capable. It connects a set to digital signal sources and usually takes the form of three red, green, and blue RCA-type jacks. It delivers video signals in three components (hence the name) consisting of a brightness signal and two signals representing color (or, more specifically, color difference—one color minus another). Be warned that not all component video jacks support HDTV. Some support only 480i signals. There are sets with one of each type—so check out the true capabilities of component video jacks if you want to hook up two different HD-component sources. Component video is being phased out of some products, including Blu-ray players, because Hollywood deems it insecure.

High-end projectors or multimedia monitors may include **RGB** jacks for connection to video processors or PCs. RGB jacks separate the signal into red, green, and blue and may use either screw-on BNC-type connectors or the more common RCA connectors. **RGB+HV** jacks add two more connections for horizontal and vertical sync. **VGA** allows connection of PCs.

S-video input, the second highest-quality analog video connection, uses a delicate multi-pin plug that separates the brightness and color portions of the video signal. Unlike component, it's not HD-capable. S-video is now rare in a/v products.

The third highest-quality analog connection is **composite video**, which mixes brightness and color together (hence the name composite).

Like component video—which sounds confusingly similar—composite video uses a standard quarter-inch-wide RCA-type plug with yellow color coding and a fat center pin. It is not HD-capable. It is used to connect VCRs, other legacy signal sources, and iOS devices.

All TVs have at least one **RF**, or antenna, input. Instantly recognizable by its large threaded jack, this connection carries multiple channels of video and audio. RF is used most often for antenna, cable, or satellite input—if you have more than one of those, you may want more than one RF input—and is sometimes labeled "75 Ohms" (really old analog TVs may have a 300 Ohm antenna input instead). You'll also see that big RF screw terminal on DVRs, VCRs, cable boxes, and satellite receivers because, like your TV, each of these products includes a tuner (channel selector). If your DTV has no ATSC tuner, then you'll need a separate tuner box. If your DTV has a **CableCARD** or **Tru2Way** card slot, ask your local cable operator for the card, and then you won't need a cable box. Otherwise, connect your cable or satellite box to the DTV using the highest-quality output, preferably HDMI or component video.

Whenever you see **video** or **audio/video** inputs listed with no other description, you can usually assume they're composite video and stereo analog audio. **Front-panel jacks**—convenient for connecting a camcorder or videogame machine—are usually either HDMI, a combination of composite video and analog audio, or both. A set with surround or stereo capability also will have **digital optical/coaxial inputs** and left/right-channel **analog audio/video inputs. Audio/video outputs** feed picture and sound from the TV to other components though HDMI-ARC is better for that purpose.

Projectors and other high-end equipment are also likely to have a multi-pin **RS-232** jack, to interact with a touchscreen interface, and a **12-volt trigger**, to send control signals to other components such as projection screens.

A **network connection** allows access to broadband-related network features. It can be wired (ethernet) or wireless (wi-fi). While ethernet provides the most reliable connection for video streaming, there is a workable wireless alternative: Upgrade your router to 802.11ac and use an **AC bridge** to get the signal from the router to your system.

A new **WiDi** standard from Intel allows a computer to stream directly to a TV without using a home network or internet connection.

The thorny subject of digital rights management

Any discussion of how sources connect to a DTV would be incomplete without mention of what you're *allowed* to connect—and why.

The term **copy protection** begs the question of who is being protected—certainly not the consumer. There are a number of more straightforward synonyms including **copy prevention**, **copyright protection**, or just plain old **anti-copying**. The latest buzzphrase is **digital rights management**.

Congress first regulated digital media with the No Electronic Theft act of 1992, which mandated the use of the **Serial Copy Management System** (**SCMS**) in component CD-R decks (but not computer drives). SCMS (pronounced "scums") allows one generation of digital copying—the copy itself may not be copied. The No Electronic Theft Act of 1997 criminalized file sharing (including the nonprofit kind). The Digital Millennium Copyright Act of 1998 made it a crime to tamper with copy-prevention schemes or to sell devices for that purpose. There are exceptions for libraries and researchers but the DMCA remains controversial among civil libertarians. Coming up are more draconian laws that would mandate the use of copy-prevention technology. In addition, the Supreme Court, in the now infamous Grokster Decision, has ruled that technology developers may be penalized for intent to cause copyright infringement, paving the way for a legal free for all.

A key battleground in digital rights management is the **broadcast flag**. The broadcast flag's stated use is to prevent video from being shared on the internet, though it may limit consumer recording and networking options in other ways. The Federal Communications Commission approved it in 2004, but a federal court ruled against it in 2005, and the FCC finally killed it in 2011. However, this bad idea may come back.

These different formats impose various levels of restriction. Following, for the sake of background, is a theoretical overview of digital rights management from least restrictive to most restrictive.

- Copy freely
- Copy once
- Copy never, use only
- Use only with registration key
- Use only in approved regions
- Use with down-resolution
- Use never

Copy freely. Those who oppose any kind of copy prevention believe this is the natural order of the universe. Some products still work this way including certain CD-R drives, MP3 players, and cassette decks.

Copy once. The leading example is SCMS. Found in CD-R decks (but not PC drives), SCMS allows you to make a digital clone. However, you cannot digitally copy the copy. The limitation is built into the blank disc as well as the player—that's why black-box CD-R decks require special "music" discs that cost slightly more than ordinary "data" discs.

Copy never, use only. Successful examples include the Advanced Access Content System, used in Blu-ray; the Content Scrambling System (CSS), which ensures that only licensed DVD players can play DVDs; and the Macrovision APS (Analog Protection System), widely applied in VHS and DVD movie releases. VCR makers cooperated with Macrovision by altering their products to work more effectively with APS. The process works by manipulating video signal levels and is designed to prevent only copying, not normal viewing, though it plagues certain video displays with picture bending or cycles of darkening and brightening. In a masterpiece of obfuscation, APS-treated software is sometimes labeled "Macrovision Quality Assurance"!

Use only with registration key. For example, a Windows PC logs onto Microsoft's server to authenticate your use of Windows. This happens once, then you're home free. Registration keys also figure in some CD copy-prevention schemes such as key2audio and MediaCloQ. These are rarely used now—most CDs are in the clear.

Use only in approved regions. Regional coding is built into most DVD-Video players and programs. It prevents releases in foreign markets from being illegally imported into other markets.

Use with down-resolution (or **down-res**). This alarming restriction would be on quality, not content. It would degrade the resolution of high-definition programming, rendering it standard-definition, through the component video outputs. Blu-ray carries the seeds of down-res through a digital flag called the **image constraint token**. Under the AACS DRM used in Blu-ray, licensed players made after December 31, 2010 must limit analog video outputs to standard-def. Players made after December 31, 2013 may not have any analog outputs at all. This is sometimes referred to as the **analog sunset**. This kind of stuff makes videophiles and audiophiles see red.

Use never. The leading example is the signal scrambling used by cable systems for premium channels and pay-per-view events. Of course you have no right to enjoy what you haven't paid for.

To be continued

DRM has become a red-hot issue among high-end videophiles. We resent the possibility that the Supreme Court's Betamax Decision, which legitimized video copying for personal use in 1984, is being rolled back in the age of digital television by corporate-technological fiat. It is even more infuriating to see early-generation HDTV sets being rendered obsolete by down-res or by the total elimination of the component video interface from Blu-ray players and other source components.

However, without DRM, Hollywood is less likely to release its crown jewels for HDTV consumption. Hollywood needs the revenue generated by its more successful movies for the same reason I need the royalties generated by this book. Without compensation, creative work becomes impossible. The result is an impasse for DTV makers. If they don't implement some form of DRM, their DTVs won't be able to accept a copyright-protected digital cable or other signal. If they do, consumers will have fewer DTV recording options (with HDMI—none).

As always, this story is *definitely* to be continued.

Shopping for a DTV

When buying a DTV, you should do a lot of watching, and while you're doing it, you should be aware of limits—the limits of different kinds of displays, and the limits of what's being fed into them. Here, then, are some pointers for the shopper/survivalist.

Screen size and viewing distance

How big a screen do you need for home theater? Though viewers may prefer varying degrees of intensity, choosing the right screen size is not just a question of taste. The size and layout of the room should figure in your decision.

The traditional formula for relating screen size to viewing distance uses screen heights. Granted, this does not make for easy calculation, since manufacturers and retailers quote screen diagonal, not screen height. Anyway, measure the viewing distance from your favorite seat to the front of the screen. Screen height should be one-third of the viewing distance for digital TV (high-definition or line-doubled) and about one-fifth for analog TV. Or viewing distance should be three times screen height for digital sets or five times screen height for analog sets.

To calculate true picture height and width from the diagonal measurement, use this formula: In a 16:9 screen, height is 0.49 times the diagonal, and width is 0.87 times the diagonal. Minimum viewing distance should be at least 1.5 times the diagonal. Or the diagonal should be no more than two-thirds of the viewing distance. To second-guess these numbers, there are numerous viewing-distance calculators online using SMPTE, THX, or other standards. Search "viewing distance calculator."

These are *minimum* distances. Some viewers may find them fatiguing and may need to sit farther back. On the other hand, if you have poor eyesight—or you're willing to trade a clean picture for emotional impact—you may be tempted to go for something larger, but be careful. When buying a DTV, you don't want to see the dotted **pixels**, or picture elements, that make up the picture. The best size is the one whose flaws are subjectively minimal when viewed at your preferred distance.

So get out your tape measure, figure out the viewing distance, and take that number to the store. Don't leave home without it! Any set you buy should look good from exactly that distance.

If you want a prospective TV purchase to fit into a piece of furniture, such as a home entertainment center, ask your tape measure to tell you whether the set's total height, width, and depth will fit. (And don't forget ventilation and the needs of a set with side-mounted speakers.)

Know what you're seeing

Don't assume the pictures you see in a store are identical to those you'll see at home. For one thing, manufacturers ship sets, and retailers display them, with radically heightened **contrast and brightness** settings. At home, you should set them lower, using the **movie mode**, both to adjust for less intense home lighting and to extend your set's life. Lower settings will take longer to wear out an LCD's backlight, plasma panel, projector lamp, or other components with limited lifespans.

As discussed in the chapter on "DTV by the numbers," the DTV standard embraces several **display formats**, including ultra-high-def, high-def, and lower-resolution formats. Any digital display will convert incoming signals to its own native resolution. Try to get a sense of how a set looks when displaying, say, 1080i HDTV, versus 480p DVD. Some digital broadcasts and cable channels are standard-definition, so how an

56

HDTV set converts and displays non-HD signals is significant. Can you see a difference between UHD and HD on a UHDTV? On smaller screens, maybe not.

A really good showroom uses top-drawer **video sources** such as a Blu-ray player for HD, or either a Blu-ray player or hard-drive-based player for UHD. This tells you what a prospective set looks like at its best. A mediocre showroom may be simply feeding every set on the floor with standard-def cable. This tells you nothing at all.

Go ahead, watch some standard-def 480i video. A digital set will use a line doubler or a more sophisticated video scaler to upconvert it to its native resolution, such as 1080p (HD) or 2160p (UHD). Many scalers are not very good, so you may see varying amounts of **motion artifacts**—distortion of moving objects or objects covered in a camera move. Look at what happens to such objects, especially those with diagonal edges. The variation between sets, given the same material, can be telling.

Try some familiar Blu-ray discs or DVDs, preferably with a player that has a progressive-scan output. Some sets will display DVD in 480p. Others will upconvert it to 540p, 1080i, 1080p, or 2160p, so you'll see more processing, but fewer scan lines. This kind of **scaling** is about to become part of your life, so learn what it looks like. The original DVD format is not high-definition television (HDTV), but it is a form of digital television with standard-definition resolution (SDTV). You may be watching a lot of purchased or rented DVDs with your new DTV.

If you're watching a UHD or HD source, find out whether it's 2160p, 1080p, 1080i, or 720p. Then find out whether, and how, the set is converting it. Watch a little of everything to build your visual experience.

Limits of picture and sound sources

One of the biggest mistakes you could make would be to buy a big HDTV while ignoring what goes into it. The best signal source available is the **Blu-ray** format for HD and, starting around now, UHD.

Over-the-air **HD broadcasts** are an excellent source of HDTV (with **UHD broadcasts** possibly starting a few years from now). Stations offering DTV channels deliver varying amounts of HDTV; the rest is SDTV. Analog TV, which survives in some cable systems, does not look nearly as good. To get broadcast TV, you'll need an antenna (either outdoor or indoor) and a tuner-equipped set. To get DTV, you'll specifically need a DTV tuner.

Satellite systems are by definition digital. However, many satellite channels are standard-definition, not high-definition. It may even be rather poor SDTV—heavily compressed, full of motion artifacts—due to the compulsion to jam as many channels as possible into the limited bandwidth afforded by satellites in orbit. Launching new satellites is, needless to say, expensive. You'll also need a slightly larger dish and an HDTV-compatible satellite receiver. Some of this may slow you down but don't let it stop you.

If you're dependent on **cable** service, ask a customer service rep for HDTV service (and an HD-capable convertor/descrambler). Also, be warned that there may be some confusion between HDTV service (using the 1080i or 720p formats) and DTV service (which may merely be digitized lower-resolution television). UHD cable is in the works.

The latest arrivals on the video-delivery scene are the telephone companies—**telcos** for short. **Verizon** is starting to offer **FiOS TV** in some (lucky) metro areas. FiOS stands for **fiber optics**, the technology Verizon is relying on to deliver the highest possible bandwidth, and therefore the highest potential picture quality. Areas not served by Verizon might get telco-TV service from **AT&T**. Its **U-Verse** uses a more economical combination of fiber and last-mile copper. Both companies are offering **triple play packages** combining television, internet, and phone service—in direct competition with cable operators. Their customer bases are on the rise even as cable and satellite dwindle.

Video streaming via the internet can offer a cornucopia of programming. And some of it purports to be HD or even UHD. However, video data rates tend to be pokey, resulting in a crude, blocky picture. That's a bad medium for judging picture quality.

Types of video noise

Part of being a videophile (or at least an informed consumer) involves knowing when a display is doing a poor job and when it's merely the victim of a poor signal. That's why it's important to recognize different kinds of video noise, motion artifacts, and other distortions.

- block noise, macro blocking
- chroma noise
- contouring, posterization
- cross-color distortion, moiré

- dot crawl
- image lag, motion smear
- image retention, burn-in
- jaggies, stairstepping
- mosquito noise
- phosphor lag
- snow
- uniformity problems
- white crush

Block noise, or **macro blocking**, causes images to dissolve into squares of varying sizes—the worse the noise, the larger and fewer the blocks. It results from compression errors in the video signal and is usually not a fault of the display.

Chroma noise adds grain to broad areas of color. Analog source signals, or those delivered through composite video, have this problem.

Contouring, or **posterization**, roughens transitions from light to dark. It is often the fault of the display, especially plasmas.

Cross-color interference, or **moiré**, is a rainbow effect that results from color information being mixed with brightness information. It's usually seen in composite video connections. It should not appear in S-video, component video, HDMI, or other video interfaces when they are properly implemented—unfortunately, that's not always a safe bet.

Dot crawl causes zipper-like or checkerboard distortion along edges, especially vertical edges, and colored objects. It's a particular problem with analog video formats and composite video connections.

Image lag or **motion smear** is a problem with liquid-crystal-based panels and projectors. The crystals move when stimulated by current, but subside slowly. To minimize this problem, look for sets with a minimum 120 Hz refresh rate—in the panel itself, all other claims are meaningless—and smoothing (or anti-judder) circuitry.

Image retention or **burn-in** occurs when graphics and other static images leave a visible impression on plasma panels and tube-based displays. Mild cases, at least in plasmas, may be reversed by running full-motion video for 24 hours. Severe cases are a permanent problem.

Jaggies cause diagonal lines to break into steps. That's why **stairstepping** is another name for this effect. The problem may be inherent in the signal or may be a fault of the display. Better displays use more sophisticated video processing circuits to eliminate this effect. The best way to gauge stairstepping is with a wind-tossed American flag.

Mosquito noise blurs the outlines of sharp objects with shimmering noise. Like block noise, it's usually an artifact of video compression (not the display).

Phosphor lag causes the outline of a bright object to persist on-screen after it's supposed to have disappeared. It happens with plasmas and direct-view sets.

Snow is also called video noise though brightness noise or **luminance noise** would be more accurate. It is a grainy impurity embedded in the brightness portion of the video signal. A bad display may look noisy, but far more often, this is a fault in the signal.

Uniformity problems occur in LED-backlit LCD sets because of the way the backlights are deployed, notably with edge lighting. Full-screen white or bright color is uneven across the surface and in corners.

White crush occurs on displays with inadequate high-frequency bandwidth. It's usually not a signal-related problem.

Comb filtering

During the analog era, one of the most important video-processing circuits in better TVs was the **comb filter**, and even in DTVs, it has an impact on analog signals (which quietly survive in libraries and gear). Its function is to reduce video distortions that occur in NTSC, the analog TV standard, due to a kind of shotgun wedding. For greater efficiency in use of the over-the-air spectrum, the color (**chrominance**) part of the signal is interleaved with, and therefore interferes with, the portion governing brightness (**luminance**). Among the most objectionable flaws is **moiré**, often visible on a herringbone tweed or a football referee's jersey. Another one is **dot crawl,** that annoying effect that mars the edges of objects (especially on sports or weather graphics). The best comb filter is none at all. That's why component video and S-video are superior forms of analog video connection—there's no need to separate color from brightness because they're not mixed together in the first place.

A digital video scorecard

In the digital television era, there is a pecking order in picture quality. Home theater buffs with the biggest screens have the biggest motivation to pay more for a better picture. After all, a screen big enough to capture your imagination is also merciless enough to magnify every flaw in the picture source that feeds it.

Following is a scorecard of video formats in order of resolution. Since half of these formats are digital and half are analog, the methods of specifying resolution vary. Digital formats usually specify resolution in a simple grid of pixels, vertical by horizontal. For analog formats, the first number is still vertical resolution, but the second is lines of horizontal resolution, or TV lines (TVL). For a refresher on this knotty subject, reread "DTV by the numbers/Interpreting TV specs."

4K or **Ultra HDTV** (2160 x 3840 pixels, progressive): This is the highest resolution supported in consumer-level video displays with more than eight million pixels (four times the number of 1080p) and more than two thousand lines of vertical resolution. That's the number of horizontal scan lines that can be counted vertically from screen-top to screen-bottom, though the whole point of HDTV and UHDTV is that they're so small, they're hard to see. UHDTVs are now common at the top ends of most lines. UHD is being used for movie production and archiving. Blu-ray has gone UHD and Comcast, the Dish Network, and DirecTV have announced plans for UHD channels. UHD streaming is available from Netflix, Amazon, and YouTube though possibly not at the highest level of quality. UHD performance is considerably enhanced by HDR technologies.

1080p HDTV or Full HD (1080 x 1920 pixels, progressive): This dominant DTV format has more than two million pixels and more than a thousand lines of vertical resolution. Like UHD, above, this progressive scanning format (that's the "p" in 1080p) scans the whole picture in one pass, like your computer monitor. Unfortunately, the 1080p signal is too big to travel through a standard-sized 6 MHz broadcast channel using currently available MPEG-2 video compression. Most HDTVs now convert all incoming signals to a native resolution of 1080p as a picture-enhancement strategy. True 1080p signals are available via Blu-ray disc, satellite, cable, telco, and streaming (some may be migrating from 1080i). However, there are no 1080p over-the-air broadcasts.

1080i HDTV (1080 x 1920 pixels, interlaced): This was once a TV format and is still a broadcast format which HDTVs can handle. It uses interlacing (that's the *i* in 1080i) to scan the picture in two interlocking passes. This may lead to distortions of moving objects called motion artifacts though the picture can still look great. This form of HDTV is available from networks including CBS and NBC and carried by cable operators and telcos. It was also offered by the Dish Network and, in slightly adulterated form, DirecTV before they adopted 1080p. Whether you opt for broadcast, satellite, cable, or telco HDTV delivery, you'll

61

need both a set top box to decode the signal and an HDTV to display it.

768p HDTV (768 x 1366 pixels, progressive): Used in some LCD HDTVs, though there is no 768p signal format per se in television. All incoming signals are converted to the set's native resolution.

720p HDTV (720 x 1280 pixels, progressive): Favored by ABC and Fox—and also available via broadcast, satellite, cable, and telco—this form of HDTV has fewer scan lines (you guessed, 720) but does use PC-friendly progressive scanning. In many DTVs, the 720p signal is up-converted or downconverted depending on the set's native resolution.

480p EDTV/DVD (480 x 720 pixels, progressive, 540 TVL): Even at enhanced-definition resolution, digital video can look great. The best way to see it is on DVD—preferably feeding a progressively scanned signal from a high-end DVD player to a DTV. Thus avoiding interlacing, the videophile gets eye-popping color (especially red) and a very clean picture, even though it's not as sharp as HDTV. Though the 1080i and 720p HDTV formats are dominant in broadcasting, some digital broadcasts are in the 480p EDTV format. While DVD has 720 horizontal pixels, if you use the old analog-TV yardstick, lines of horizontal resolution (TVL) are 540 or less.

480i SDTV/DVD (480 x 720 pixels, interlaced, 540 TVL or less): Low-end DVD players deliver an interlaced picture. The interlaced (480i) format is a step down from the cleaner picture possible with progressive DVD (480p). However, it still beats analog broadcast, cable, and VHS sources hands down, and probably will outperform most satellite signals. (See "Picture & Sound Sources/Disc players" for more on the difference between 480p progressive and 480i interlaced DVD.)

480i SDTV/satellite (480 x 704 pixels, interlaced, 480 TVL or less): Non-HD satellite broadcasting has the same number of lines as DVD. However, most satellite receivers use interlacing, diminishing picture quality to about the level of good analog television. By stuffing their systems with as many channels as they can, satellite operators further reduce picture quality, leading to blocky pixellated distortion. This has done little to diminish the popularity of satellite service though it also probably has fed much of the enthusiasm for DVD. Lines of horizontal resolution (TVL) measure at 480 or less.

An analog video scorecard

At this point we move from digital video, measured in vertical by horizontal pixels, to analog video, where horizontal resolution is measured only in TV lines (TVL). All of these formats are interlaced.

Laserdisc (480 x 425 TVL): The 12-inch precursor to DVD delivered nearly the same resolution as its five-inch replacement. Unfortunately, Pioneer stopped making LD players in 2009, though you can still buy them used.

Super VHS (480 x 400 TVL): While S-VHS delivers a picture with slightly more than 400 TV lines, approaching the resolution of laserdisc, its picture is full of noise. That's because it excels over VHS only in the recording of the brightness (luminance) portion of the video signal—the color (chrominance) portion is no better than regular VHS.

Cable (480 x 300 TVL): Cable operators can and do deliver HDTV and SDTV. Cable operators are phasing out analog channels. They want to reclaim the bandwidth and are willing to provide set top boxes that accommodate ancient analog TVs.

VHS (480 x 250 TVL, interlaced) **and D-VHS** (up to 1080 x 1920 pixels, progressive): VHS is both the worst and the best of all possible worlds. It is limited to 250 TV lines with a softening of the picture that's easily visible even on a 20-inch set. But if you're one of the lucky few with a digital D-VHS recorder or HD DVR, you can record HDTV.

Surround Sound

*A*fter the big picture, big sound is the other major requirement of a home theater system. Like a big screen, surround sound is designed to engulf the senses, suspend disbelief, and pull the audience into the story. In a good home theater, the quality of surround sound can easily surpass anything available in your local cineplex. You have more control over the level of sound, which is abusively loud in today's moviehouses, and the quieter moments of a movie will have greater impact without the senseless yapping of today's horrendously rude cineplex audiences.

Surround speakers

Speakers are among the most important components in any system, whether for movies or music, in surround or stereo. And sometimes the bulkiest. Human hearing being variable, they're perceived in different ways by different listeners. That's not just a matter of specs and acoustics—musical tastes and hearing capacity also play a role (the young hear better). However, once you've found a set you like, they'll serve you for at least a decade (or as long as the drivers and the parts surrounding them hold out). So invest in the right speakers.

A typical surround sound installation includes a **5.1-channel** (or more) array of speakers driven by a receiver. As explained way back in the introductory chapter, that usually includes three speakers in front (left/center/right) and two toward the rear of the side walls (left/right) plus a subwoofer (or two). Some systems have three front speakers, two side-surrounds, and one or two back-surrounds in a **6.1-** or **7.1-channel** configuration though in my opinion few rooms need that much surround coverage. Height or width channels, some derived from processing modes, have substituted for (or supplemented) back-surrounds.

The advent of Dolby Atmos and DTS:X makes height speakers more significant than they originally were. That's because the height effects in these formats are not merely derived from processing—now there's real spatial information for those height speakers to reproduce. This has brought a new three-digit nomenclature with the final digit representing the height speakers. So a 5.1-channel system with height speakers above the front left and right is a **5.1.2-channel** system. One with height speakers in all four corners of the room is a **5.1.4-channel** system. With both height and back-surround speakers, it would be a **7.1.2-channel** or **7.1.4-channel** system.

Within that basic multichannel array are many possibilities. You might anchor the system with large front left and right speakers and dispense with the sub. Those left and right speakers may have built-in subs of their own. You might have five identical satellites and a subwoofer, or four identical sats, a horizontal center design, and a sub. You might prefer your center speaker to be a horizontal design or a duplicate of the front left and right. Purists prefer the center speaker to be a perfect duplicate of the other front speakers. If a center speaker isn't convenient, you might eliminate it entirely and run the surround processor in phantom center mode. You might prefer bipolar or dipolar speakers in the rear. You might have one, two, or no back-surrounds, or two to four height speakers, or two width speakers. There are many variations.

Regardless of what you choose, it's important to match the front and rear speakers to produce a solid **soundfield**. Even if their enclosures are not the same size, try to match their driver sizes and types, so that they'll all deliver **directional** information (front/back, left/right) in a uniform way. That's called **timbre matching** and it generally points toward a one-brand speaker system. However, don't assume all speakers from the same brand have the same tweeter or woofer or crossover or voicing. Check the specs—and listen.

The larger the room, the bigger your speakers should be. Their job is usually described as "moving air"—a physical action—though it would be more precise to say that speakers create changing patterns in air pressure like ripples in water. Bigger drivers in larger enclosures can do more of this. A bigger room also requires more bass energy, so bigger speakers might be worth the investment in space and money.

Main left/right speakers

Within every surround system is a vestige of the old stereo pair. Front left/right speakers can take many forms. They may be the same size or larger than center and surround (rear) speakers, especially for listeners who prefer using large speakers to play in stereo. For a uniform surround soundfield, it's generally best for the center speaker to resemble the others to the extent possible, sharing tweeter sizes and materials. Drivers that handle midrange/woofer duty should be the same or very similar in construction.

Floorstanding speakers (or **tower speakers**) are usually big enough to produce serious bass. You don't need big speakers to put together a good home theater system, especially if the system includes a subwoofer; many speaker makers even recommend that you cut off low bass going to the main speakers and route it entirely through the sub. But many audiophiles still prefer the old method of using two bigger speakers, both for music reproduction, and for better bass coverage in general. You might consider buying big main speakers *instead* of a subwoofer (surround 5.0), setting the surround processor to route all bass through the front left/right speakers. Look for slimline designs with big woofers limited to the side of the enclosures. Add internal amplification to power those side woofers and you've moved into the next category.

Powered towers are extremely practical. First, they make it easy to eliminate the separate subwoofer from the system—your spouse will like that. As a bonus, they add *two* built-in powered subwoofers to your system. *You'll* like that! More benefits: The designer will make sure that the subwoofers blend well with midrange and high-frequency drivers. And with the speakers' internal amplifiers driving the subwoofers, your receiver may be less likely to distort, giving the whole system more dramatic dynamics and a sense of ease. Only a few speaker models use this configuration, but for bass addicts, it may be the best way to go. Keep in mind that you'll need to run not only a speaker cable to each speaker, but a power cord as well, to feed the internal amp.

Bookshelf speakers (or **near-field monitors**, to use the recording studio term) are also an excellent foundation for a home theater system. The name is somewhat deceptive—most speakers aren't designed to sound their best on shelves, where vents can be blocked, and acoustic conditions are poor. But many *bookshelf-sized* speakers have just enough

Clockwise from top left: The PSB Synchrony Series, including the Synchrony Two shown ($3000/pair), offers high quality of construction including seven-layer side panels and extruded-aluminum corners. Cambridge Audio's Minx S215 ($800) is a compact 5.1-channel satellite/subwoofer set whose tiny cube speakers use 2.25-inch flat diaphragms to achieve surprisingly good sound, with wide dispersion, though due to their size, the sub has to produce more of the bass frequencies. The Paradigm Seismic 110 subwoofer ($1500) can be set up with the Perfect Bass Kit ($299) to tailor bass response to the room, eliminating bass-bloating humps. Finally, the Klipsch RP-140SA ($500/pair) is a Dolby Atmos-enabled speaker with top-firing drivers that handle the height channels in Atmos and DTS:X. It is voiced a little "bright," which enables height effects to stand out more.

67

bass to function without a subwoofer, so you can save your pennies and buy a really good subwoofer later to firm up the bottom octave. They also happen to sound excellent in stereo. The speakers in my 5.1-channel and 2.1-channel systems fall into this category.

LCR speakers typically have a woofer-tweeter-woofer array and can be used for all channels, or just the front three channels, or just as a horizontal center. Note that the dual-woofer array in a horizontal speaker can cause a sum-and-cancellation effect called **lobing**, audible as uneven midrange and bass response depending on where you sit. To avoid this, some designs add a midrange driver to anchor voices. A few others vary the crossover circuitry between the woofers and tweeter.

Satellite/subwoofer (or **sat/sub**) **sets** replace big speakers with small satellite speakers that get a boost in the lower octaves from a subwoofer, which is usually **active** (self powered). They're a popular way to introduce surround sound into a room that hasn't got space for boxy conventional speakers. Depending on their size and bass response, the sats may or may not function independently of the sub. That's fine since most surround receivers let *you* decide how bass is divided between sats and sub. The smallest sat/sub sets function best in a small to medium-sized room. Because they lean heavily on the sub—which must produce a wider range of frequencies to compensate for the small size of the sats—there may be some unevenness where the bass meets the midrange, indicating poor integration of sats and sub. However, the best mini-sat/sub sets can and do sound good.

Dolby Atmos-enabled speakers, which provide height effects in that new surround standard, take two forms. They can be **integrated speakers**, floorstanding or monitor, with top-firing height drivers. Or they can be top-firing **add-ons** to existing floorstanding or monitor speakers. Speaker layouts that work for Atmos also work for **DTS:X**. Dolby Atmos and DTS:X are discussed in more depth in the chapter on "Understanding surround standards."

The center speaker

The front center speaker—the anchor of a 5.1-channel or more surround system—is not without its controversies. Everyone agrees that the center channel is where dialogue resides and that it plays a significant role in delivering special effects and—with concert videos, high-res music formats, or stereo sources adapted to surround—even music. After that, opinions diverge.

Some high-end surround buffs insist that nothing less than three identical front speakers will provide the best results, pointing to acoustic problems associated with center-speaker designs using a woofer-tweeter-woofer configuration. In some horizontal centers, as noted above under "LCR speakers," the two woofers may cancel each other out at certain frequencies. However, many speaker makers and consumers are equally comfortable with (or resigned to) horizontal front center speakers.

Surround and back-surround speakers

The term **surround speakers** refers, somewhat confusingly, to the rear speakers (as opposed to the whole set) because they provide the surround effects. In 6.1- and 7.1-channel systems, there are surround speakers against both the side and back walls.

The main choice in surround speakers is between **direct radiating** (read: conventional) speakers and the trendy **bipole/dipole** types that have drivers on two sides of the enclosure. The front and rear drivers in **bipolar** speakers operate **in phase** with one another—that means they move in the same direction. **Dipolar** speakers operate **out of phase**. Some speakers can be switched to operate in either mode. A few can also switch off one set of drivers to function as **monopole** (conventional) speakers. Especially with dipoles, the main effect is a more subtle and diffuse feel in the surround effects. The drivers fire to the front and rear of the room, not at the listening position, to keep surround effects from becoming an unwelcome distraction.

In 5.1-channel systems, a single set of surrounds operates on the back of the side walls and they are referred to, simply, as the surrounds. In 5.1+ systems, they are called **side-surrounds**, while two additional speakers on the back wall are called **back-surrounds**. A 7.1-channel system uses two back-surrounds; a 6.1-channel system uses only one. The addition of back-surrounds is an idea that made sense in public movie theaters, where it originated as a way of spreading surround effects evenly through the rear of the auditorium, but then it was hastily ported to home theater without any consideration for practicality. The 7.1-channel configuration is usually overkill unless your room is a cavern.

In 6.1- and 7.1-channel systems, THX recommends bipole/dipole speakers in the side-surround positions and direct-radiating speakers in the back-surround positions.

69

Height and width speakers

Dolby Pro Logic IIz, **Audyssey DSX**, and **DTS Neo:X** introduced the concept of height channels, while DSX and Neo:X also introduced width channels. Height speakers are hung above and outside the
main front left and right speakers. Width speakers go just outside the
main left and right speakers. Because these new listening modes deliver
fairly amorphous information, small satellite speakers will suffice for
them. They are explained in "Understanding surround standards."

While height channels were initially greeted with a yawn, the concept has taken on new life with the advent of **Dolby Atmos** and **DTS:X**
in home surround gear. In these new standards, height and floor speakers combine to create a three-dimensional soundfield in, around, and
above the seating area. Loudspeakers are starting to accommodate
height channels by building top-firing drivers into the tops of **Atmos-
enabled speakers**, though dedicated ceiling speakers can be used in a
more elaborate home theater installation. See "Understanding surround
standards."

The subwoofer

In either surround or stereo, a subwoofer (or two) can make a big
difference. **Driver size** may be 8, 10, or 12 inches (occasionally more).

A subwoofer may be a simple device that uses a single 10-inch driver, something economical in terms of both space and money. Or it
might be a large heavy object with multiple **active drivers** and/or **passive radiators**, the first powered by internal amplification, and the second by the powerful movement of air within the enclosure.

Smaller subs are more suitable for most apartment dwellers. With
rare exceptions, a driver size lower than eight inches disqualifies the device as a subwoofer, regardless of what the manufacturer calls it, and
qualifies it as a woofer. Such un-subwoofers are seen mainly in low-end
compact systems.

At the other extreme is the driver that's so big that the amp has
trouble starting and stopping its movement. It doesn't help that the
struggling amp may be deceptively specified at its peak (rather than continuous) power rating, which is legal, because amps built into subwoofers aren't covered by Federal Trade Commission regulations. Even so, a
large room can benefit from the muscle of a well-designed 12- to 15-
inch sub. Just don't get excited and buy a bad one.

When in doubt, buy a simple 10- to 12-inch model with at least one driver facing you and nothing more than a thin speaker grille in between. That setup will deliver low bass in roughly the same time domain as the other frequencies, keeping the sound in focus. Such **front firing** designs produce the most accurate bass, though that dry description hardly does justice to the indescribable feeling of big bass waves hitting your body. **Bottom firing** designs achieve much of the same effect.

Subwoofer designs are diverse enough that drivers may face any direction. In fact, they may sit entirely within the enclosure, emitting sound through a **vent** or **port**. There might even be tubing involved. These **bandpass subwoofers** are efficient and may come with impressive specifications. Unfortunately, listeners pay the price in muddier and less accurate bass.

Most subs come with both a **high-pass filter** and a **low-pass filter**. The high-pass filter is generally used with a speaker-level connection (i.e. speaker cables) while the low-pass filter is generally used with a line-level connection (interconnect cables). The high-pass filter is sometimes fixed at a certain crossover frequency, or may be affected by the crossover control. It enables the sub to strip out the lows and pass the rest of the signal to the satellites. The low-pass filter is variable, adjustable via the sub's crossover control, and limits the sub to sounds below the crossover frequency. A surround processor contains a low-pass filter of its own that duplicates the one in the sub.

For that reason, a desirable feature found in select subs is a **low-pass filter bypass**. It eliminates the sub's low-pass filter and lets the receiver or surround processor determine what bass frequencies go to the sub. By eliminating the sub's filter when it's not needed, the bypass results in cleaner and more accurate bass. Sometimes an input labeled **LFE** (low frequency effects) does the same, bypassing the sub's filter.

Subs with servo circuits can track and control woofer movement with great precision. Infinity invented this kind of sub amp. (I have fond memories of an Infinity HPS sub that unnerved people on the opposite side of the testing-room wall. They came in and asked, "Are you OK?") Paradigm, Velodyne, and a few other brands offer something similar.

A rare but highly desirable feature in subwoofers is **equalization**. An equalizer is a glorified tone control. In a sub it smooths out frequency-response humps that occur when bass waves interact with the room. EQ may be either **graphic** (fixed frequency bands) or **parametric** (variable frequency bands). The kind built into subs is usually parametric. Some subs with EQ come with a kit that allows you to measure the room and dial in the best settings. Infinity's RABOS kit uses a sound meter and test CD allowing the user to plot a chart, identify bass humps, and adjust the sub to neutralize them. Paradigm's Perfect Bass Kit uses a stand-mounted microphone, computer software, and USB connections to computerize the process. ELAC's Auto EQ runs on an app-driven mobile device. Any of these, done right, can work wonders.

In-wall and in-ceiling speakers

It's possible that you—or a loved one—may not want to live with a home theater system that resembles Easter Island. If your spouse resents having all those obelisks cluttering up the room, in-wall or in-ceiling speakers may make the difference between having a surround system and not having one. The best in-walls are high performers. Here are just a few things you'll need to know.

In-wall speakers are as subject to the laws of acoustics as any other kind of speaker. It's even more important to get the placement right because you're making a hole in the wall or ceiling. For that reason, use a well-qualified custom installer to get it right. (See the chapter on "Connecting a Home Theater System/Hiring a custom installer.")

To provide a little wiggle room, there are some in-wall designs that use a pivoting tweeter, which can be moved around to change the dispersion pattern. However, woe be to the consumer whose in-walls just aren't in the right places.

In-ceiling speakers suffer the acoustic disadvantage of having to fire at the floor. In bare-floored rooms this becomes especially problematic. While you wouldn't want to use them for front speakers in a surround system, they might be a helpful problem solver in the rear positions. They can also be used to implement Dolby Atmos and DTS:X. Look for in-ceiling models with angled, pivoting, or motorized drivers.

For best acoustic results—especially with an in-wall sub—you'll need to install a **back box** behind the speaker to control resonance. Otherwise the wall becomes a resonating chamber. Result: bad sound.

On-wall speakers

The popularity of flat-panel video displays has spurred the demand for flat-panel speakers that wall-mount with **keyholes** or **threaded inserts**. Some sound better than others but all operate under a handicap.

Simply placing conventional speaker drivers into a flat enclosure is not a recipe for good sound. The enclosure is a critical component of a speaker, and without enough physical depth, a speaker cannot produce much sonic depth.

Wall placement brings further acoustic problems. The closer the speaker is to the wall surface, the higher in frequency (and more objectionable) its interaction with the wall becomes.

Alternatives to on-walls include in-walls, which don't interact with the wall surface at all, because they're behind it, *in* the wall. Also consider small satellite speakers on slender stands—they can be both visually nonintrusive and great-sounding.

Soundbars

This speaker category aims to complement flat-panel displays. Soundbars may be **active** (with built-in amps) or **passive** (without amps). They may support two, three, five, 5.1, or 7.1 channels. One fairly simple type is a three-channel speaker that handles the front left, center, and right channels, leaving the surround and sub channels to other speakers. Another type usually tries to simulate surround sound from a single horizontal bar, sometimes with external sub. It may use either legitimate Dolby/DTS surround processing and/or various faux-surround modes. A small but growing number have HDMI jacks that support higher-quality DTS-HD Master Audio and Dolby TrueHD surround. Wireless connectivity can be also helpful, either for the sub connection, or to connect Bluetooth mobile devices. Although some high-end models stand out from the pack—from Polk, MartinLogan, Phase Technology, and Sony—most bar speakers are more suitable for a bedroom system than for a primary home theater system.

Soundbases

As an alternative to soundbars, some manufacturers offer soundbase speakers. The main difference between a soundbar and a soundbase is that soundbase has a wider top surface and the TV sits on it. There are not many good-sounding soundbases. Zvox is one of the few manufacturers that does a good job in this category.

Exotic speaker types

While speaker technology is staid and stodgy, there are a few alter-

73

native technologies to liven things up. One is the **horned speaker**, which places the tweeter deep into the enclosure, in a recess designed to shape the flow of sound in a specific pattern. This maximizes loudness in the listening positions covered by the horn's dispersion pattern. That in turn makes horned speakers the most efficient type, allowing your surround receiver to achieve higher volumes during peak musical and cinematic moments with less strain. It also introduces cupped-hands coloration, though an experienced designer can minimize this problem. The acknowledged masters in this area are Klipsch—whose founder, Paul Klipsch, is remembered as the father of the horned speaker—and JBL, which has been selling large pro-level horned speakers to movie theaters since the dawn of the talkies. In conventional speakers, tweeters are often slightly recessed into **waveguides**, distant cousins of horns.

Electrostatic speakers are a legitimately high-end flat-panel speaker technology. Electrostats replace conventional cone- or dome-shaped speaker drivers with a diaphragm that operates in a charged field. This eliminates any need for sound-polluting crossover circuitry between drivers. And it provides swooningly realistic high frequencies and imaging—the increased clarity can be both refreshing and addicting. However, low frequencies don't fare as well, so most electrostats are best mated with a subwoofer to beef up the bass. Also, some electrostatic models require a direct power connection, so if your room doesn't have enough AC outlets, add your local electrician to the list of ancillary expenses. Quad and MartinLogan are famous for their electrostats.

Another form of flat-panel speaker technology hails from England where **Hi-Wave Technologies** (formerly **NXT**) has pioneered the use of flat membranes—made of various plastics and even glass—in lieu of conventional woofers and tweeters. The panels are stiff, but light, and excited by a simple motor-like device just as you can excite a tabletop by rapping your knuckles on it. Eliminating conventional cones and domes provides more even coverage. And again, there's no need for a crossover. There are some sonic disadvantages too: The panel may add coloration of its own. And bass is on the lightweight side. Cambridge Audio uses HiWave technology in its Minx sat/sub speaker system.

Flat diaphragms stimulated in the conventional fashion—by electromagnetic voice coils—are also appearing in speakers from the Harman International stable (Infinity, JBL, Revel). Flat diaphragms used as tweeters are called **ribbon tweeters**. A **pleated-diaphragm tweeter** figures in speaker lines from GoldenEar, MartinLogan,, and others.

Wireless speakers, currently used in a few soundbars and HTiB systems, are getting a leg up with the new **WiSA** standard from the Wireless Speaker & Audio Association, a group of speaker makers including Aperion, Definitive Technology, Klipsch, MartinLogan, Paradigm, and Polk. WiSA uses the uncluttered 5.2 to 5.8 GHz band to deliver uncompressed 24-bit, 96 kHz audio—which is very high quality—with robust error correction and 24-channel switching.

Interpreting speaker specs

- drivers
- voice coil
- enclosure: vented or sealed
- two-way, three-way
- driver types: woofer, midrange, tweeter
- driver materials
- driver sizes
- crossover
- power handling
- impedance
- sensitivity, efficiency
- frequency response
- dispersion
- shielding
- matching center, surround

The best way to buy a set of loudspeakers is to take over the store, throw out all the staff, acoustically treat the listening room to eliminate all echoes, bring in your own reference amps, and plot response curves from test signals. In real life, the best way to find good speakers is to listen to a lot of them before buying (or get a money-back guarantee from manufacturers who specialize in online sales). Before you go shopping, it doesn't hurt to learn a little about what speaker specs mean. You can look up many of these specs on the websites of speaker makers.

A loudspeaker is a box containing one or more **drivers** (paper, plastic, metal, or composite **cones** or **domes** in metal **baskets**) that emit different frequencies of sound when stimulated by electrical signals. An electromagnet called the **voice coil**, suspended in a magnetic field created by other magnets, is connected to each driver and controls its movement. **Crossover** circuitry determines which frequencies, low or mid or high, go to which drivers. The drivers "move air" in the room: they cre-

75

ate the variations in air pressure that we perceive as sound. Now, how does that translate into specs?

Start with the **enclosure**. As a practical matter, is it too big for the room? The type variously known as **vented, ported,** or **bass reflex** contains a hole in the front or back to leak a little more bass into the room. Slightly rarer is the **sealed** or **acoustic suspension** type, which may be more accurate if the designer has a good ear. All other things being equal, a vented design will play deeper, but watch out for muddy-sounding peaks in the midbass response, as well as audible port turbulence (**chuffing**). A well-made speaker will have a **braced enclosure,** with structures inside the enclosure that prevent it from resonating. The enclosure itself should vibrate as little as possible—unwanted resonance adds undesirable coloration to the music, especially in upper bass. A good enclosure does not have a note of its own; when you rap on it with your knuckles, you should hear a dull thud, not a discernible pitch.

In a **two-way** design, the drivers consist of a **woofer**, which reproduces lower sounds, and a **tweeter**, which reproduces higher sounds. **Three-way** designs add a **midrange** driver and a second crossover. More drivers may or may not produce better sound—again, it's up to the designer. Rare **coaxial** speakers, notably from KEF and ELAC, place the woofer and tweeter in the same axis, one inside the other. That produces better imaging, and more accurate sound over a wider area, because the listener is always the same distance from both drivers. A possible downside is cupped-hands coloration though good design can overcome it.

Stick to **woofer sizes** greater than 5 inches if you want to avoid total dependence on the subwoofer for bass response. Avoid woofers greater than 10 inches unless they're built into a subwoofer or mounted on the side of a tower.

Driver materials vary in the maximum speed at which they vibrate, something particularly critical in the tweeter. The fastest materials are metal (aluminum or titanium), though good-sounding drivers may also be made of plastic (usually polypropylene), textiles (including silk dome tweeters and bulletproof Kevlar woofers), paper (often treated or "doped" with something to enhance sound or durability), or various composites and laminates (like the aluminum/ceramic composite made and invented by Infinity). Faster tweeters are not always better. Misused, metal tweeters can be harsh and ringy. Paper, the cheapest material, can sound excellent. Also pay attention to the driver surrounds (which fasten the driver into the basket) and to the basket itself (diecast metal is better than pressed metal).

Debating the fine points of **crossover** design is beyond the scope of this book. But be advised that while there are excellent speakers with crossovers between the midrange/woofer and tweeter at anywhere from 1.8 kHz to 4 kHz, 2 kHz is believed to be a significant number. Avoid tweeters crossed over below 2 kHz (they're likely to distort) or 8- to 10-inch woofers crossed over above 2 kHz (they're likely to create a cupped-hands effect with voices).

Manufacturers may specify **power handling** as a **range**, a **peak** number, or an **average** number. Does that mean lower-powered amps are easier on your speakers? Not necessarily. A struggling amp may generate harsh distortion that hurts both your speakers and your ears. More power is better than less. Clean power is better than dirty power. This will come up again when you buy a receiver or power amp.

Your receiver or power amp, or its manual, contains clues to what kind of speakers you need. Speakers are rated for **impedance**, the amount of impediment a speaker puts up to an audio signal entering through its terminals. Think of impedance as a faucet: the more you open the faucet, the more water runs through. Likewise, as impedance drops, the speaker's demand for current rises, and the amplifier's workload goes up. In speakers, an average impedance is 8 Ohms, which is easy for any amplifier to drive. Impedance of 6 Ohms is a little harder, 4 Ohms harder still, and 2 Ohms or less requires high-end heroics. Speakers operate at a range of impedances that varies according to frequency—in other words, a loud, percussive bass drum thud might have a lower impedance than a whispery flute solo, and therefore would exert more strain on the receiver. Many inexpensive receivers will not comfortably drive anything much lower than 8 Ohms, so examine this spec carefully when mating any speakers with any amplifier.

Sensitivity is a crucial spec which may be given confusingly. Specified in **dB** (**decibels**), usually using a one-watt test tone measured one meter from the speaker, it tells you how loud the speaker will play in a non-echoing (**anechoic**) test chamber. An alternative version of this spec is **room efficiency**, which substitutes an average room for the non-echoing chamber. When you see that term, allow for rating inflation of 2-3 dB. Otherwise consider 88 dB about average. Lower sensitivity figures are forgivable in high-end speakers, assuming you have powerful amps to drive them. Higher ones suggest that the speakers will provide extra volume capability, though possibly at the cost of accuracy or listening comfort. Sensitivity directly affects how much power your system needs to perform well. Never forget that that every 3 dB cut in sensitivity requires a doubling of amplifier power to achieve the same volume.

Frequency response attempts to capture the speaker's distribution of low, middle, and high sounds in a few numbers that may be helpful or highly misleading. The compact disc format is designed with a frequency response of 20Hz-20kHz (20 Hertz to 20,000 Hertz, or vibrations per second). Human hearing starts lower but rarely goes as high. If a speaker measures at 40Hz-25kHz +/-3 dB, it has pretty good bass, high-frequency response well beyond that of the CD, and does all that with reasonable consistency—within 3 dB. Within 6 dB or more is meaningless; within 1 or 2 is very good. Of course, notches or peaks in different parts of the frequency-response curve will result in different sound quality, and less scrupulous manufacturers will use that as an opportunity to lie like crazy with this spec.

It's the manufacturers who accurately specify **dispersion** who restore our faith in human nature. What they are doing is specifying frequency response as measured directly in front of the speaker, and then a few degrees off-axis, giving you both figures to suggest how the sound changes as you move out of the **sweet spot**. That's the acoustically perfect spot in which you are always supposed to sit and, in real life, cannot always sit. The more degrees off-axis a speaker can go, while maintaining uniform frequency response, the more consistent that speaker will sound as you move around the room.

Finally, when buying any speakers for surround sound use, be sure to find out if **matching** center and surround speakers are available, and in what configurations. Part of the job of researching any speaker purchase is to investigate the entire line that interests you. Find out what the options are, what mates best with what, and look for matching driver sizes and materials.

Connections

- 5-way binding-post speaker terminals
- wire-clip terminals (spring-loaded or clamp-down)
- dual speaker terminals for biamping/biwiring

Most audio buffs agree that screw-down binding posts make better speaker terminals than wire clips. Depending on your definition of practicality, you might like the **5-way binding post** for its versatility. It can accept various kinds of terminated cable: **Spade lugs** are U-shaped connectors that fit around the post in a binding post. **Banana plugs** flex (or the good ones do) and fit into a hole in the post. **Dual banana plugs** are built into plastic casings that facilitate connection. **Pin connectors** are slim rods that fit into the hole in the side of a binding post or under the tab in a wire-clip terminal. Bare wire works with everything—to prevent oxidation over longterm use, you may solder the exposed copper, though that isn't strictly necessary to achieve a viable connection.

Some speakers use **wire clips**. The cheap version allows the user to simply lift a plastic tab and stick in the speaker-cable tip. There's also a sturdier cylindrical type becoming increasingly popular in satellite speakers—it's all-metal and more durable. Wire clips on speakers are usually spring-loaded. They work best with bare wire though they'll also accept fancy cable terminated with slender pin connectors (which certainly do not sound better than bare wire, though they're more durable).

Some speakers have an extra pair of terminals to allow **biamplification** (separate amp channels feed each driver) or **biwiring** (separate cables feed each driver, joined at the amp). Biamping provides the most dramatic benefit, assuming the extra amp channels are available. For systems not using biamping or biwiring, the dual terminals can be bridged, and the bridges are usually supplied.

Shopping

Not everyone has local access to a retail listening room with a broad selection and good acoustics. In fact, the decline of a/v specialty retailing has led several brands to adopt an internet-dominant strategy. Buy-before-you-try makes you dependent on reviewers, though a money-back guarantee and free return shipping can reduce the risk.

If one's available, your local high-end emporium is a good place to start listening even if much of what it offers is beyond your budget. At least you'll get an idea of what good sound is like. And you might discover something wonderful that you'd never have found in a large chain store or over the internet.

Here's an absolute must: *bring your own program material!* Familiar music will tell you more about a speaker than will a salesperson.

Listening conditions may contain snares. The most common sales ploy is to play two competing models at slightly different volumes. Because human ears hear more bass and treble at higher volumes, the speakers played louder will always sound better, even if the difference in volume is so slight as to leave the listener unaware of it. You can't be

79

sure if your little receiver will perform as well as a high-end system's muscle amps and pristine pre-pros. There are many ways to be misled.

Give yourself a chance to hear how the bass response of a powered tower (or other floorstanding speaker) differs from that of a satellite/subwoofer set where the sub provides all the bass. Compare them both to a system based on bookshelf-sized speakers—looks dull, performs well, lots of manufacturers excel in this size range.

If you and/or the sales staff are more ambitious, ask if you may hear the same speakers—something you like—fed by a budget receiver, a middling one, a top-line model, and a separate configuration of preamp-processor and power amp. With or without subwoofer. Or with more than one subwoofer. If you'd really like to stir up trouble, ask: "Um, do speaker cables really make a difference?"

Surround receivers & components

A fancy TV will provide a wealth of switching possibilities but its built-in sound system is rarely strong enough to suspend disbelief. Basing a home theater system solely on a big TV is no more practical than building a mega-sound system to accompany a 13-inch bedroom set. A serious home theater sound system will be based either on a surround receiver (the best bet for most people) or on a combination of surround preamp-processor and power amplifier(s).

Receiver or separates?

- audio/video receiver
- preamp-processor
- multichannel power amplifier

An **audio/video receiver** (or surround receiver) combines the functions of power amp, preamp (including volume control), surround processor, video processor, and audio/video switcher into one box. That saves money, space on the rack, and eliminates the mass of cables needed to keep an amp and pre-pro on speaking terms—it's the essence of practicality. The downside is that it's hard (though not impossible) for delicate surround processing circuits and beefy amp components to coexist in the same enclosure. When they share the same power supply, the amp tends to draw power away from the pre-pro at peak moments. But in a good design, it all works. Over the years manufacturers have conquered some pretty serious design challenges to make the a/v receiver a practical and even high-performing product.

There is, however, a higher-end alternative to receivers. Separating the **preamp-processor** from the **multichannel power amplifier** (or rack of amps) allows both to operate at a more high-end level of performance, with more purity for the pre-pro, and more power for the amp. At least, that's the theory. In the real world, mating an inadequate amp with a feature-poor pre-pro can easily result in lower performance and less versatility than a well-designed receiver could provide at the same price. A high-end buff who knows the ropes can get the best performance from separate surround components, but the most practical reason to go for separates would be to fill a large room, something a muscle amp can do more easily than most receivers. If you're not sure whether separates are for you, buy a receiver with 5.1-channel (or more) analog line outputs. In a later upgrade, you'll be able to use those outputs to feed the inputs of a multichannel power amp. Still further down the upgrade path, you might consider replacing the receiver with a proper pre-pro.

If you want to integrate an existing stereo system into a surround

The Denon AVR-X7200WA ($2,999) has the nine amp channels necessary to run Dolby Atmos and DTS:X with a full complement of four height speakers. It currently serves as the author's reference receiver for use in loudspeaker reviews.

system, get a surround pre-pro with **analog pass-through** for your stereo preamp. If you love your old stereo power amp enough to keep it in your home theater system, a **three-channel amp** will provide the extra amp channels needed for the front-center and surround channels. However, I don't recommend mixing amps. Some amps will react differently than others when you adjust the master-volume control, making it harder to keep all channels tracking up and down in a balanced manner.

Interpreting receiver/amp specs

- power rating (watts per channel)
- total harmonic distortion
- analog direct
- classes of amplification

The power rating for a typical receiver or power amp might read: "100 watts per channel, RMS, 20Hz-20kHz, into 8 Ohms, at .05 percent THD." That means the amp delivers 100 watts to each speaker, with continuous power, at full frequency response, assuming the speaker's nominal impedance is 8 Ohms, with fairly low distortion. **RMS** (or root mean square) refers to the method of measuring current. (See "Surround speakers/Interpreting speaker specs" for more about impedance.) The Federal Trade Commission mandated what are now called **FTC power ratings** decades ago following a scandal in which manufacturers were found to be playing fast and loose with their specs (and the truth).

Some power ratings are measured using a **1 kHz** (one kilohertz) test tone. That provides another stable frame of reference—even though a one-note test tone really has nothing to do with music, which swells and falls and contains pitches ranging from low to high.

Some specs mention **peak power**, measuring the amp's performance at a loud, transient high point. With subwoofer amps, which are not separate products unto themselves and thus exempt from FTC rules, this still happens routinely. Take all peak amp specs with a grain of salt.

FTC notwithstanding, manufacturers still cheat on ratings. One trick is to narrow the range of frequency response at the bass end to eliminate the demands of low bass reproduction. Another involves driving only one or two channels at a time, giving the power supply a break (an "all channels driven" spec is more valid). Yet another way to achieve a higher watts-per-channel spec is to specify an impedance lower than 8 Ohms (in other words, a speaker that accepts more current). As the impedance drops, the number of watts goes up. However, a manufacturer that specifies power ratings into 8 Ohms and at lower impedances (say, 6 Ohms) is providing useful information. Given side-by-side, the two specs will show you how the amp behaves when it feeds a more demanding speaker. Two different models may both deliver 100 watts into 8 Ohms, but if one delivers 120 watts into 6 Ohms while the other delivers 150 watts into 6 Ohms, the latter model delivers more current.

All other things being equal, more power is better. It makes your system play louder, sound better, and less likely to fry the speakers with harsh distortion. However significant the power rating may be, it would be a mistake to assume that amps with the same power ratings are identical. A good one is more likely to have a heavier power supply and higher current capability (check weight specs). To drive demanding low-impedance speakers, investigate outboard multichannel amps.

Total harmonic distortion (THD) measures the amount by which an amp alters the waveform of a presumably pristine movie soundtrack or piece of music. It's generally assumed that distortion is harsh and nasty, although it can also sound mellow in vacuum-tube amps. Tiny differences in THD—say, .01 versus .05 percent—are rarely detectible by human ears. Some say differences of up to 1 percent are insignificant. Don't get caught up in the numbers game; it's the *nature* of distortion that determines whether, and how much, it's acceptable. Truly nasty distortion affects **higher-order harmonics**. (Harmonics are the higher tones that are sounded when a note is struck. They are multiples of the original tone. They're what enable you to distinguish middle C played on a guitar from middle C played on a piano.) The precise nature of distortion is something you won't find in the specs. Trust your ears.

If you have a high-quality analog signal source in your system, such as a turntable/phono-preamp combo, check if a receiver has an **analog direct** (or **analog bypass** or **pure direct**) mode. Why? Because the average receiver translates all analog signals to digital by default. Audiophiles prefer to switch analog signals in the analog domain when possible. Analog direct mode is most useful with full-range speakers; smaller satellites with limited bass response require bass management and a sub, which is usually handled digitally.

Most surround receivers and power amps use **Class AB** analog amplification. Class AB is a hybrid of two other types of analog amplifier. **Class A**, occasionally used in high-end stereo systems, keeps both of its power stages full of current, which wastes a lot of power in the form of heat—though it can sound great. **Class B** pushes and pulls current between the two power stages, so that one is always on and the other off, which is more efficient, but not as sweet-sounding. Class AB keeps current in both stages part of the time but alternates between them, providing the ideal compromise between sound quality, efficiency, and cost.

Class D amplification is finding its way into a growing number of receivers and compact systems. These products convert the analog input signal to a train of pulses that is amplified by a rapidly switching output stage which is always either on or off. The amplified pulse train is low-pass filtered to recover the analog waveform and eliminate ultrasonic switching noise. This process dissipates far less energy in the form of heat than conventional all-analog amplification and allows products to be made much smaller, slimmer, and lighter—and far more energy-efficient. Originally the sound quality of Class D lagged behind its elegant premise. But the technical hurdles are being overcome and some Class D products now sound excellent. Someday the majority of surround amps may be made this way. (Incidentally, the D in Class D does not stand for digital, and some manufacturers of Class D products object to them being described as digital amps.) Makers of Class D receivers include Pioneer, which calls its version D3, and Rotel. One form of Class D technology, **ICEpower**, is licensed to various manufacturers from Bang & Olufsen.

Other alternative amplifier topologies include **Class G** and **Class H**. These analog amps shift from low-voltage rails to high-voltage rails in much the same way that a car shifts gears. This allows greater power output with same-size parts and/or greater energy efficiency. The difference between Class G and Class H lies in the rails—Class G uses multiple rails, while Class H uses variable rails—though the terms may also be used interchangeably. Arcam uses Class G in some products, while AudioControl uses Class H in some products.

Controlling your receiver

A receiver has many functions. So how you operate it affects what you'll get out of the product and how you'll feel about using it. There are several ways to control a receiver: through front-panel buttons and knobs, through an onscreen interface combined with a remote control, through a smartphone or tablet app, or occasionally through a web browser interface.

Front-panel controls are usually rudimentary. Generally they'll let you power up the unit, select a source input, and select a listening mode.

Some add menu navigation and other functions behind a drop-down door, duplicating much of what's on the remote control.

But for the most complete command of settings, including basic setup, you'll use the **graphic user interface**. It is almost always intricate—but it offers a good overview of what your receiver does and how it works. It's easier to fool around in a GUI than to read a manual. An above-average GUI makes adroit use of color and diagrams to clarify functions. A select number use **context-sensitive help**, using additional text to explain functions and controls as you go along. This can be tremendously helpful and saves you from diving into the manual.

To operate the GUI and basic functions such as volume and mute from a seated position, you'll be at the mercy of the **remote control**. Does it have control codes for your other components **preprogrammed** into it? Is it capable of **learning** codes for other manufacturers' products? Look for backlit or glow-in-the-dark keys if you plan to use the remote in a traditional darkened home theater. Some manufacturers are simplifying their receiver remotes and that can be helpful to those intimidated by receivers. However, be warned that the simplest-looking remotes are not necessarily easier for the more advanced user. See "Useful Accessories/Remote controls" for more pointers.

More and more receivers support free **control apps** that allow the **iPhone, iPad, or iPod touch** to serve as the remote control. Apps for **Android** devices are now equally common. Yamaha has a control app for Amazon's Kindle Fire. If you've been using the mobile device all day, you may like using it at night as well.

A few models also support a **PC and web interface**. The receiver has an IP address that can be entered into a PC's web browser. Once the browser recognizes the receiver, you can operate some settings and functions. This is an especially good way to operate multi-zone features when you're not in the same room as the receiver.

Ease of use

Ease of use is a critical and often overlooked aspect of buying a receiver. This isn't just another set-top box or disc player—this is the heart of your system, and you're going to be using it a lot. Do the control menus make sense to you? On the remote, are there too many buttons? Or too few, making some oft-used functions less accessible? Are controls well differentiated by size, shape, color, labeling, or layout?

One thing that's making receivers much easier to use is **auto setup**

and calibration. It places a small microphone—supplied with the receiver—in the listening position to gauge variables like speaker size, speaker distance, and acoustic conditions. The receiver then adjusts output levels for each channel and various other settings. If you haven't had much experience in setting up surround systems, auto setup will make your introduction to surround more painless and help you get good performance out of your new system immediately. As time goes on you might become curious enough to buy a sound pressure level (SPL) meter, measure the volume level coming out of each speaker, and fine-tune your receiver's settings manually. But for most beginners, auto setup is the way to go.

Another benefit of auto setup is that it corrects acoustic problems (see next section on "Room correction").

Room correction

Receivers and preamp-processors with auto setup also have the ability to correct problems with room acoustics. Good room correction can do a lot to improve performance, especially bass response. But bad room correction can be worse than none at all. Whether it helps depends partly on the quality of room correction and partly on the room.

Every room—except for the non-echoing chambers where manufacturers test speakers—has **resonant** or **room modes** that cause **peaks** and **nulls**. Even a speaker designed to provide a relatively flat frequency response under ideal testing conditions will sound louder at certain frequencies at home. **Standing waves** are stationary patterns of high and low volume caused by waves bouncing between opposite walls. The frequency of the standing wave depends on the distance between walls. Effectively, the room becomes a resonator. Bloated bass is a typical standing-wave problem.

These problems can worsen if certain room dimensions are identical or are multiples of one another. If two pairs of surfaces have a common resonance, the problem is even harder to correct. Wall construction can also contribute to resonance problems. For instance, a single layer of scantily reinforced sheetrock may resonate and muddy the room's sound considerably. That's why fancy multilayered sheetrock is becoming chic in high-end custom-installed home theaters.

Room correction measures room modes from one or more listening positions using a supplied setup microphone. Depending on the product, the correction may involve only one or two peaks, or inverse equalization may be applied across the whole frequency spectrum. Better room correction schemes work not just in the frequency domain, but in the time domain as well, using **phase control** to compensate for time-domain errors. Getting them in phase results in a more coherent sound.

One way to apply room correction is to use a separate device called an **equalizer.** Equalizers can be **graphic** (adjusting fixed frequency bands that can't be changed) or **parametric** (with the ability to select the affected bands) or both. Setting them up often involves the use of simple microphones and a **real time analyzer**. EQs suitable for home theater use and real time analyzers are available from AudioControl and Gold Line, among others. THX-certified EQs are available from Rane.

But it's far more convenient to buy a surround receiver or preamp-processor that has room correction built in and performs it cleanly in the digital domain. Some products use technologies licensed from **Audyssey** or **Trinnov**, while others use their own proprietary technologies. For more about Audyssey, see "Understanding Surround Standards/Audyssey's auto-setup and listening modes." Most major manufacturers of surround receivers, and some pre-pro makers, include auto set-up and room correction. Some ELAC, Infinity, Mordaunt-Short, and Paradigm subwoofers also include room correction specifically for bass.

To do room correction right, you must place the setup mic at ear level. The easiest way to do this is with a camera tripod. Brands that include a mic stand with receivers include Anthem, Denon, and Marantz.

Room correction is not without its detractors. For one thing, they say, correcting for one listening position doesn't help others—a universal one-size-fits-all fix is not possible. That's why most room correction systems now measure from more than one position to avoid favoring a single sweet spot. Poorly done room correction may also cause audible problems such as skewed frequency response, phase errors, or ringing. Even so, minimal and carefully done room correction may help correct gross acoustic problems, especially with the interaction between subwoofers and room modes.

Video features of receivers and pre-pros

Video-related features of surround receivers and preamp-processors are not limited to video pass-through and switching. Receivers have gotten slicker at the way they handle video.

One helpful feature is **video upconversion**, the ability to convert between video formats so that signals entering the receiver through the lower-quality composite and S-video inputs can exit the receiver, en route to the video display, through the higher-quality HDMI or component video outputs. That doesn't improve the quality of the signals, but does make your system easier to set up, because you need to connect only the highest-quality video output to the display.

If you trust the HD or UHD video processing in your HDTV or Blu-ray player, the receiver should have **1080p pass-through** or **UHD/4K pass-through** to refrain from mangling the signal as it passes through. Note that there is a difference between 1080p or UHD/4K pass-through (the receiver passes the signal from source components to TV without alteration) and **1080p upscaling** or **UHD/4K upscaling** (the receiver upscales video to a higher-res format). Ultra HD compatible receivers with 4K pass-through may or may not have 4K upscaling. Similarly, **progressive-scan video processing** converts the half-frames of an interlaced (480i) signal to the full frames of a progressive-scan (480p) format, smoothing out jagged edges and other video artifacts. Some receivers have marquee video processors licensed from the likes of Anchor Bay and Faroudja though most now use no-name processors.

Video processing circuits (in the display and disc player as well as the receiver) can cause a lip-sync problem onscreen because they take so much time to swallow up and manipulate whole frames of video. When the video lags the audio by a few seconds, this causes a disconnect between the voices you hear coming out of the speakers and the lips you see moving onscreen. An average person can detect as little as a single film frame's worth of delay—people who transfer films to video can detect as little as a half-frame. Many receivers offer **lip-sync delay** to delay the audio so the video can catch up. Some offer a **game mode** that bypasses the video circuits to keep fast-moving visuals in sync with audio.

Wireless and network audio features

Wirelessly streaming from smartphones and tablets has become an indispensable feature. There are two kinds. A receiver may have one, or both, or neither.

Apple AirPlay allows an iPhone, iPad, or iPod touch to turn the receiver on, select the right input, and start streaming audio from the iDevice. AirPlay requires an ethernet or wi-fi home network connection. Some USB jacks also allow a wired connection to Apple devices.

The alternative to AirPlay is **Bluetooth**, a direct device-to-device link that does not go through the home network. Bluetooth is found in many phones, computers, and other devices. It has a dizzying array of profiles at various levels of quality and functionality. Bluetooth compresses the signal before sending it. It can function with **aptX** or **AAC**, compression methods that offer higher audio quality. Sony's version is **LDAC**. **Bluetooth NFC** (near field control) compatibility allows the two devices to sync with a simple bump. In some receivers, Bluetooth requires an extra-cost adapter; in others it is built in.

Some receivers are certified by the **Digital Living Network Alliance**. **DLNA**-certified receivers communicate with a PC's hard drive through the router and the Windows Media Player to access music, video, or photos. DLNA also certifies TVs, Blu-ray players, PCs, cameras, and mobile devices. While DLNA is usually implemented via ethernet, it is also **Wi-Fi Direct** certified. The latest version of DLNA is 4.0, introduced in 2016. It solves the "media format not supported" problem by mandating transcoding on the media server.

As alternatives to AM and FM, receivers can include network-connected **audio streaming apps** for services such as Pandora and Spotify as well as internet versions of radio stations. These are becoming more common even as satellite radio is waning. Some receivers include the internet version of SiriusXM as opposed to the satellite version, which has become rare.

To use AirPlay, DLNA, or streaming, your receiver needs to connect to your home network by **ethernet** or **wi-fi**. Receivers with Bluetooth, AirPlay, and wi-fi are especially convenient. Sony, Denon, and Onkyo are examples of what I like to call **triple threat receivers**. Note that you can't count on Bluetooth, AirPlay, and wi-fi being baked in. Often one or more will be supported with an extra-cost adapter.

Introduction to receiver connections

"There comes a time in the affairs of man," W.C. Fields once said, "when he must take the bull by the tail and face the situation." Yes, it's time to think about what jacks you need on your receiver's back panel. Count your components. Then consider whether the receiver has enough inputs and outputs of the right kind to suit your system's current and future needs. (For the sake of simplicity, this section will refer only to receivers, but the information applies to pre-pros as well. You may also refer to the "Connection glossary" under "Connecting a Home Theater System" and to chapters on other products.)

Highest-quality video and audio connections

- HDMI digital interface
- MHL interface
- HDBaseT interface
- IEEE 1394 digital interface
- component video inputs/outputs
- multichannel analog line inputs and outputs

HDMI, MHL, HDBaseT, and IEEE 1394 are highest-quality digital video interfaces on receivers. HDMI is by far the dominant one. It is contributing to the long-desired uncluttering of receiver back panels. When these digital connections are not available, the next best options are their analog equivalents, component video and 5.1- to 7.1-channel analog audio.

The most desirable jack on the back of a receiver is **HDMI**. It carries both video and surround (or sometimes just stereo) audio. It's available in multiple versions from 1.0 to 2.0a. Don't assume that the latest version is supported in every receiver. See the "DTV connections" chapter for a complete rundown on HDMI's multiple versions. Get 1.3 or higher for the best sound or 2.0a for the most up-to-date UHD/HDR and 3D video support.

A note on connecting an SACD player to a receiver: The best method is a direct DSD signal via HDMI. The second best method is to set the player to output high-res PCM via HDMI. The third best method is the multichannel analog interface, bypassing HDMI.

MHL (Mobile High-Definition Link) is a new audio/video interface designed to patch a smartphone into an HDTV or home theater system. It has a compact 5- to 11-pin plug at one end (usually, but not limited to, mini-USB) and an HDMI plug at the other. It carries 1080p video and 7.1-channel audio, both uncompressed, and allows the TV to charge the mobile device. Onkyo and Pioneer, among others, support it.

HDBaseT is a rare but welcome HDMI extension interface that uses Cat 5e or Cat 6 ethernet cable for high-quality video and audio. It is more robust for long cable runs than regular HDMI.

One use for **IEEE 1394**, mainly on older Denon and Pioneer products, is for high-resolution audio signals from DVD-Audio and SACD players. However, 1394 is also capable of carrying video and for that reason appears in camcorders.

Component video inputs/outputs are next in the videophile pecking order. Like HDMI, they're HD-capable, albeit analog. Component connections are bulky—using a trio of red, green, and blue RCA-type plugs—and inadequate bandwidth can kill their advantages. According to the vigilant videophiles of the Imaging Science Foundation, few receivers or preamp-processors provide adequate bandwidth for component video connections, so you might be better off making component video connections directly between your DTV and source components. At least one receiver maker, Cambridge Audio, is eliminating component video. Others may follow, since the interface has been phased out of new Blu-ray players.

Receivers with **multichannel analog line inputs** can be fed by disc players with **multichannel analog line outputs**. This arrangement came about in the early days of Dolby Digital, when there were still many receivers that didn't support it. Nowadays, with HDMI passing most formats, those multichannel-ins are less useful. When a disc player can convert high-res audio signals to PCM via HDMI, that is a better form of connection than multichannel analog, which is why multichannel-ins are disappearing from receivers at lower price points. But they still serve Blu-ray players with high-quality digital-to-analog conversion via analog-out such as the higher-end Oppos. Because several surround formats are 7.1-channel, multichannel analog line outputs usually add the extra pair of back-surrounds to the basic 5.1-channel array.

A receiver with **multichannel analog line outputs** can act as a surround preamp-processor when plugged into a multichannel amp. Again, multi-outs are disappearing from receivers at lower price points.

Among the receiver's 7.1-outs is at least one **subwoofer output**, called **LFE (low frequency effects)**. It is usually a single RCA-type jack. A receiver with two sub-outs can feed two subs. A sub may have stereo inputs, but feeding just one of them is fine—the sub doesn't need a stereo signal. (The other method of connecting a sub is to go from the front left/right speaker outputs, through the sub's speaker-level ins and outs, to the speakers. This will make use of the sub's **high pass** filter—in other words, the sub will strip out the low sounds for its own use and send the rest of the signal to the main speakers. A few subs support high-pass filtering via line-level connections.)

Other video connections

- S-video inputs/outputs
- composite video inputs/outputs

- front-panel audio/video inputs
- DVR/VCR inputs/outputs
- monitor outputs
- multi-zone outputs

After component video (above), the round multi-pin **S-video** is the second best analog input/output. Unlike composite video (below) it separates the brightness and color parts of the signal to reduce video artifacts such as dot crawl and moiré. Most receivers now omit S-video.

All a/v receivers have **composite** video inputs and outputs color-coded with a yellow RCA plug. Nearly all products that provide any kind of legacy video connection have composite video. It is inferior to S-video and component video because it mixes the brightness and color portions of the signal, leading to buzzing rainbow-like moiré distortion. Composite jacks are being reduced in receivers, but not eliminated because they still provide backward compatibility for legacy formats.

Front-panel audio/video inputs are typically analog stereo audio occasionally with composite video or (preferably) HDMI. Front inputs let you plug in a videogame, camcorder, or other source component.

DVR or VCR inputs/outputs mate the audio ins/outs with composite video (always) and S-video (sometimes).

Monitor outputs let the receiver send a video signal to the TV. Be sure any receiver you buy has a monitor out that matches the highest-quality input of your video display. Today that usually means HDMI.

Multi-zone outputs serve additional displays outside the room. They may be HD-capable HDMI or non-HD composite video.

Other audio connections

- digital audio inputs/outputs (optical or coaxial)
- stereo analog line inputs
- multi-room/multi-zone, multi-source
- headphone output

Most receivers have **digital audio** connections, mostly inputs, sometimes accompanied by one or two outputs. These connections may be labeled **PCM** (which stands for **pulse code modulation**, an uncompressed digital signal) or **S/PDIF** (Sony/Philips digital interface).

There are two types of digital jacks, **optical** (also known as **Toslink**) or **coaxial**. (The term coaxial wins the confusion sweepstakes.

Strictly speaking, it refers to a type of cable construction with an outer sheath surrounding an inner core. More generally, it refers to a variety of cables with different-sized plugs and different uses, from digital cables to the RF cables used by cable TV systems.) Optical cables offer the advantage of defeating **hum**—an audible and unwanted low tone coming out of your speakers—because they pass the signal optically rather than electrically, thereby breaking up the closed circuits known as **ground loops**. Even so, some prefer the coaxial type (with RCA connectors) for its allegedly superior sound quality. Coaxial cables are sturdier than optical cables, which strangle the signal when bent. And the coaxial connection allows you to substitute inexpensive composite video cables for expensive digital audio cables, as long as the yellow-color-coded cables have a characteristic impedance of 75 Ohms.

Except for Blu-ray's lossless surround and SACD, which are best delivered by HDMI or the multichannel analog interface, digital inputs are an alternative way to connect a DVD or CD player to your receiver, though there are other options. You can also connect a disc player with 5.1- to 7.1-channel analog line outputs to a receiver or pre-pro with 5.1- to 7.1-channel analog line inputs. If you'd prefer to listen to Blu-ray discs or DVDs in stereo, you can use the player's stereo (or **downmix**) outputs, and select the two-channel option from disc and player menus. Those digital jacks are also good for plugging in CD players. If you do digital recording with a CD-R/-RW deck, it can be helpful for the receiver to have digital out as well as digital in.

Stereo analog line inputs accept CD players, cassette decks, and any device with analog outputs. If you still want to record cassettes, make sure there's a stereo analog line output as well, possibly labeled "tape loop." Most receivers provide more of these than you'll need, but if you've got a rack full of legacy formats, make sure you have enough.

Most receivers convert signals entering through their analog inputs to digital. If you're plugging in a CD player or other digital device, the signal must go through the CD player's digital-to-analog converter, then through the receiver's analog-to-digital converter. In this situation, use a digital connection between CD player and receiver to bypass those unnecessary D/A and A/D stages—you'll get cleaner sound.

Sometimes a receiver with an analog direct mode will sound better with any device that has a good-sounding analog output (including phono preamps and bleeding-edge CD players).

Higher-end receivers and pre-pros have **multi-room** (or **multizone**) **outputs**. These may be just for audio, or for both video and audio. The video connections are generally low-res composite video, though better receivers will offer second-zone HDMI. Audio connections are usually stereo, and can be analog, digital, or speaker outputs. **Multi-room/multi-source outputs** can process two different sound sources for two rooms. One family member could sit in the den watching a DVD while another sits in a nearby bedroom listening to a CD.

A receiver's **headphone output** will usually be the **1/4-inch** type that accepts what used to be called a phone plug (visualize an early 20th century Bell System operator jamming those big plugs into a patch bay). Portable devices are more likely to use the **1/8-inch** type, also known as the **mini-plug/jack**. If you want to use mini-plug headphones with your receiver, a simple 1/8- to 1/4-inch adapter will help.

Portable player connections

- iPod/iPhone/iPad dock
- USB

The **iPod/iPhone/iPad dock** is a natural extension of the receiver. Using the 30-pin docking connector on most older iOS devices, it helps the mobile devices feed your system. Now that Apple has shifted to an eight-pin Lightning connector, many docking products have become obsolete unless an awkward adapter is added. But that shouldn't affect a receiver that accepts a direct USB iOS connection. Note that USB connections don't cover all generations of iOS devices.

Some docking stations also handle video, with a composite video output to feed the receiver. They usually recharge the player (an alternative to the PC-USB link or extra-cost AC adapters). The player's screen may be disabled—but depending on the design, you can use either an onscreen display through your video monitor, front-panel receiver display, or both. An advantage is the ability to operate the iPod with the receiver's remote.

A few receivers include a **USB** jack for players inside or outside the iOS universe as well as external hard drives or thumb drives. This can be especially valuable if the USB jack accepts direct input from a computer (most don't). Check to see what audio file formats and data rates are supported by your receiver.

Radio connections

- AM/FM antenna inputs
- HD Radio
- XM or Sirius satellite radio
- ethernet

If you listen a lot to the radio, your receiver may well have **AM/FM** antenna inputs, but they lead to a barebones chip-based tuner that won't be sensitive enough to pull in far-off stations. You might either add an outboard AM/FM tuner to your system or graduate to surround separates including a good tuner/preamp. In stereo receivers, anything lacking AM/FM capability is classified as an **integrated amplifier** or something other than a receiver.

HD Radio, a digital supplement to AM and FM over-the-air broadcasting, is supported by a few receiver makers. It allows multiple channels in a single broadcast frequency.

Some surround receivers support **satellite radio**, either XM, Sirius, or both, though the number has waned. You'll need a satellite antenna, a $20 add-on, and a subscription. The SiriusXM internet radio channel has become a viable alternative and manufacturers now prefer it.

Ethernet and wi-fi connections support the various network audio features described a few pages back including internet radio.

Legacy connections

- phono input (moving-magnet or moving-coil)
- ground terminal
- tape-loop inputs/outputs (for analog recording media)
- RF-modulated Dolby Digital input (for laserdisc)

Still got a shelf or two of LPs? No serious music lover or audiophile would make fun of you for that. But your turntable will need a **phono input**. It's not an ordinary line-level input. A phono input accepts the fragile sub-line-level signals generated by tiny magnets and coils within your phono cartridge when its stylus wiggles in those mysterious black grooves. Unless indicated otherwise, most phono inputs are compatible with **moving-magnet** phono cartridges, which are relatively reliable, don't cost much, and produce good bass. However, some receivers' phono inputs are switchable for either moving magnet or **moving-coil**

cartridges, which according to audiophiles can more accurately reproduce high frequencies. Most receivers with phono inputs also provide a
ground terminal to accept another connection that defeats hum in the
turntable/cartridge. If your receiver has no phono input, or you want
higher-quality vinyl playback, you'll have to dedicate one of its line-level
inputs to a separate **phono preamp** (which might accept one or both
types of cartridge). In addition to line-level outputs, some phono preamps also have USB outputs, useful for ripping vinyl, and/or headphone outputs.

Analog **tape loop** jacks include both inputs and outputs. That not
only allows you to record and play without swapping cables—it also
helps you monitor recordings in progress when using cassette decks and
other analog recording media. Some preamp-processors provide **record
and listen** controls that allow you to record from one input while
watching/listening to another.

Out of respect for one of home video's most distinguished dead
formats, some manufacturers add an **RF-modulated digital input** to
for the small number of laserdiscs encoded in Dolby Digital.

Control connections

- IR remote jacks
- 12-volt trigger
- RS-232

Every receiver has an IR (infrared) window on the front panel to
accept remote control signals. However, if the receiver lives in a cabinet,
without a direct line of sight from seating area to front panel, remote
control signals can't reach the window. Therefore some receivers have
one or more **IR remote jacks** on their back panels to connect IR repeaters that can accept remote commands from the seating area.

To coordinate the powering up and down of components, such as
multiple amplifiers in a large-scale system, you need a **12-volt trigger**—
though some versions of HDMI also coordinate components.

RS-232, from the computer sphere, allows firmware upgrades, a
touchscreen interface, and other custom-install operations.

Speaker connections

- collared binding-post speaker terminals (UL-approved)
- wire-clip terminals (spring-loaded or clamp-down)
- A/B speaker terminals (binding post or wire clip)

As stated in the chapter on speakers, binding posts are superior to wire clips. In receivers the most common type is the **collared binding post**. It's less versatile than the five-way binding post, but does a better job of concealing wire tips from curious little fingers, reducing the likelihood of electrocuting an unwary child or pet. At least, that's what the Underwriters Laboratories believe, and for legal reasons many receiver makers (and some high-end power-amp makers) follow suit. Collared binding posts make it impossible to use spade lugs, which audiophiles prize, though they do allow banana plugs and bare wire. Binding posts provide a more secure and better-sounding connection than wire clips.

Occasionally receivers substitute **wire clips** for binding posts, at least for some channels. There are two kinds of wire clips. Though either one is perfectly capable of snapping to pieces in your hand, the **spring loaded** type is marginally more durable, and usually allows use of thicker (16-gauge or 12-gauge) wire tips. **Clamp down** wire clips are found on very low-end receivers and inexpensive compact systems and are to be treated with the same delicacy as Fabergé eggs. In terms of child safety, a wire clip is as hazardous as any binding post because it leaves bare wires exposed.

Some receivers provide two sets of **speaker outputs** labeled **A** and **B** for the front left and right channels. The B set parallels the A set and allows connection of a second set of stereo speakers.

Most receivers come with seven channels of amplification. The last two channels can be configured for a variety of uses including back-surround, height, or width speakers; or for biamplification of the front left and right speakers; or for second-zone speakers.

Shopping

A big-box or online retailer may not allow you to listen before you buy. But if you have the opportunity, grab it. If your first chance to listen is at home, treat the first listen as an audition, and if you're not satisfied that the receiver does a good job of running your speakers, get a refund or an exchange.

Contrary to myth, receivers with the same power rating do not all sound alike. Even if you've found a few models with the right power, connections, and other features, sound quality still is not a settled issue. As with any audio shopping expedition, bring your favorite listening material—Blu-ray discs, DVDs, CDs, LPs, whatever—and be prepared to listen your ears off.

If possible, audition receivers with your own speaker models. If that's not possible, find out what you can about the speakers being used for the audition. It never hurts to know what you're listening to. Pay attention to sensitivity, which governs how loud different speakers play when given the same amount of power, and impedance, which must be carefully matched when mating speakers and receivers or power amps.

The first thing to determine is how much volume the receiver can support with the connected speakers (assuming they're similar to yours). It's not unusual for a receiver to use half of its volume range to play loudly—say, 85 decibels, which you can measure with a cheap sound meter. Two-thirds of volume potential is pushing it. Beyond that point the receiver is straining—or **clipping**—as it runs out of power. Then you should move up to something more powerful.

Cycle through all the receiver's modes. Listen to it in DTS-HD Master Audio—that's how most of your Blu-ray discs will sound. Try Dolby Digital—that's how your DVD collection will sound. Listen to it in Dolby Pro Logic II—that's how your CDs will sound when rechanneled to surround. Is it too aggressive, harsh? Too vague, mushy? It's OK for a receiver lean to one side or the other, but if it does not perform well in the modes you use most often, consider another model.

Next, switch to stereo, leaving only the front left and right speakers operating. Crank up the volume. If you prefer not to use surround modes when listening to music, this is how your stereo CDs and LPs will sound. Even if you prefer listening to music in surround, stereo is a less forgiving medium than surround. That makes it a good test, and will suggest how you'll later feel about the receiver's overall performance.

By now you've spent some time with that remote in your hand. Love it? Hate it? Are the control menus easy to navigate? If not, move on to more promising models.

Never buy a receiver as an impulse decision.

Understanding surround standards

Surround sound gets more complicated with each new wrinkle. Among several organizations vying for your surround-sound dollar are **Dolby Laboratories, DTS, THX,** and **Audyssey**. These companies do not make hardware. Instead, they license their surround technologies to manufacturers. Their conflicting *and* overlapping formats form quite a thicket. The upside is that you can choose to have a system with between 5.1 and 11.1 channels. Software encoded in 5.1 sounds fine on 7.1 systems, and by the same token, 7.1 software won't trip up a 5.1 system. A 5.1.4 system will upmix non-height material to use its height channels. Diversity means more consumer choice. The downside of surround format proliferation is that it frustrates the universal consumer desires for clear, stable standards and general ease of use. If enough people are turned off by the riot of surround formats, the confusion may even slow consumer acceptance of surround sound in general. The ultimate winner of this increasingly crowded race may be our old friend stereo.

The following descriptions will move from older surround technologies to newer ones, grouped together by generation. If you want to read about the latest and greatest, skip forward a few pages.

Introduction to Dolby standards

- Dolby Pro Logic, Dolby Pro Logic II/IIx/IIz
- Dolby Digital 5.1, Dolby Digital EX
- Dolby Digital Plus, Dolby TrueHD
- Dolby Atmos

Whether your source of surround-encoded programming is Blu-ray, DVD, VHS, satellite, cable, broadcast, or streaming, Dolby encoding formats cover all these bases and more. **Dolby Pro Logic, Pro Logic II, Pro Logic IIx,** and **Pro Logic IIz** decode the oldest form of analog surround, adapt stereo sources to surround, and add extra channels to surround sources. While cineplexes use a variety of surround formats, in home theater systems the dominant player is **Dolby Digital 5.1,** available via BDs, DVDs, DTV broadcasts, satellite, cable and streaming.

99

Dolby Digital is a 5.1-channel format. Its 6.1-channel variant is **Dolby Digital EX**. Newer members of the Dolby family are the more efficient and higher-performing **Dolby Digital Plus** and the cutting-edge **Dolby TrueHD**, the latter also used on Blu-ray. The latest and most sophisticated technology is **Dolby Atmos**. All are discussed in later sections.

To learn more about Dolby surround standards, visit dolby.com.

Introduction to DTS: the un-Dolby

- DTS 5.1, DTS-ES Matrix, DTS-ES Discrete, DTS 96/24
- DTS Neo:6, DTS Neo:X
- DTS-HD High Resolution Audio, DTS-HD Master Audio
- DTS:X

Although the Dolby-licensed surround formats dominate home theater systems, they do have a competitor in **DTS Inc.**, originally **Digital Theater Systems**. Like the Dolby formats, DTS originated as a system for moviehouses, where there is more diversity in surround formats (also including, for example, **SDDS**, or Sony Dynamic Digital Sound). And like Dolby, DTS has steadily infiltrated home surround gear.

The family of DTS standards includes: **DTS 5.1**, a competitor of Dolby Digital; **DTS-ES**, a competitor of Dolby Digital EX; **DTS 96/24**, a lossy but high-resolution format used in some DVD-Audio titles; **DTS Neo:6**, a six-channel competitor of Dolby Pro Logic II and IIx; and **DTS Neo:X**, a competitor of Dolby Pro Logic IIz. The most sophisticated forms of DTS are **DTS-HD Master Audio** and **DTS-HD High Resolution Audio**, the former dominant in Blu-ray. The latest and most sophisticated form of DTS is **DTS:X**.

To learn more about DTS, visit dts.com.

Dolby Pro Logic and Pro Logic II

The oldest home surround decoding technologies exploit an older form of surround encoding. Their second function—and it's an important one—is to adapt stereo sources, like CDs, to surround.

Surround sound in the home started with **Dolby Surround**, a few years after it made a big theatrical splash in *Star Wars*. Dolby Surround's great strength is pure efficiency. It can travel in two channels, which various surround modes will expand in both front and rear. (Incidentally, Dolby also uses the old term Dolby Surround in an entirely new and un-

100

related context—referring to the receiver feature that deploys non-Atmos signals in an Atmos system. See the section on Dolby Atmos for more details.)

Though old-school Dolby Surround signals are still around—in VHS tapes and analog cable channels, among other places—the format didn't last long in home surround gear. It was replaced by **Dolby Pro Logic**, which uses the same Dolby Surround signal, but decodes it into four channels: front left, center, right, and a monaural surround channel served by two speakers, plus subwoofer. The addition of the front-center channel makes dialogue more intelligible and gives the soundfield some much-needed tightening. And of course, the sub channel enables a system to produce the monster bass home theater buffs crave.

Enabling this neat trick is a process called **matrixing** which buries sonic information for the front-center and rear-surround channels in the front left and right channels of a stereo soundtrack. Dolby Pro Logic has a forgivingly soft-focused sound that lets orchestral soundtracks breathe and makes dialogue and sound effects a little less harsh—in other words, sounds good, turn it up! Pro Logic's downside: It's a little less detailed, especially in the rear-surround speakers, which it serves with just a single monaural signal. This makes panning between the surround channels impossible. And the surround channels are truncated at 7 kHz, well within audible range.

Pro Logic never quite gels with CDs and other stereo music sources, and that's a problem. You can't just leave your receiver in Pro Logic mode all the time and expect stereo music to sound as good as surround-encoded movies. Given stereo source material, Pro Logic has a clumsy way of shoving too much of the mix into the front-center channel. The end result sounds more like mono than surround. Mindful of its flaws, surround design guru Jim Fosgate set out to achieve a cleaner sound that meshes effortlessly with any stereo source, and licensed the result to Dolby Labs.

There's more to the resulting **Dolby Pro Logic II** than the addition of a roman numeral. Dolby Pro Logic II does the following differently than the original Pro Logic: Breaks the mono surround channel into a pair of stereo surround channels. Provides more sophisticated filtering (or none at all) for surround effects. Provides user control of the center image and front/rear balance. Does all this with simpler processing, resulting in cleaner sound. And, in the most radical departure, is designed for use with stereo as well as surround-encoded material.

Pro Logic II has four modes: **music, movie, game**, and **emulation**. The music mode has a more spacious feel than the center-dominated, dialogue-oriented movie mode. Music mode sounds superb—you might prefer it over stereo for two-channel sources. Emulation is merely a reproduction of the original Pro Logic. It is inferior to the new music and movie modes and can be safely disregarded.

Pro Logic II's **dimension control** is an especially powerful tool. It works like a front/rear balance control. Two kinds of material become easier to listen to with this control at your fingertips. The first is the stereo-turned-surround mix with so much "phasey" (hollow, disembodied) material that it drives the decoder into a rear-aggressive sound. You could move much of the rear-surround material toward the front, so you feel more a part of the audience, relieving any discomfort at being too near the performers. The second is the flat, clean mix with everything miked up close and very little room ambience. You could break up the tightly focused front soundstage to let it breathe toward the rear.

The **panorama control** further heightens surround effects by leaking some information from front to rear. This deepens the soundstage, and may benefit some recordings, though for maximum realism you might prefer not to use this option.

The **center width control** distributes information among the three front channels. It can either push things inward toward the center channel, or push them outward toward the left and right channels.

There is no 5.1-channel DTS equivalent of Dolby Pro Logic II, but there is a DTS Neo:6 (see next section).

Dolby Pro Logic IIx/IIz, DTS Neo:6/Neo:X

To accommodate evolution toward 6.1- and 7.1-channel surround systems, Dolby Pro Logic has received a further update with **Dolby Pro Logic IIx**. This format expands two-channel as well as 5.1-channel source material to 6.1 or 7.1 channels, deriving the one or two added channels from the existing ones. Think of it as a retrofit for sources that have fewer than 6.1 or 7.1 channels—its purpose is to fill speakers in an expanded surround system that otherwise would lie silent. DPLIIx affects a wide range of two-channel software including videocassettes with Dolby Surround encoding, CDs, MP3s, and audiocassettes. It also works on 5.1-channel sources such as Blu-ray and DVD titles and HDTV broadcasts with Dolby Digital 5.1, DVDs with DTS 5.1, DVD-Audio, and SACD. Like Pro Logic II (minus the *x*) DPLIIx has separate modes for music and movie playback and adds a third mode for videogames.

Why did Dolby name this new format Pro Logic IIx rather than Pro Logic III? I can only speculate—and here goes. The name may reflect an ambivalence about undermining the 5.1-channel surround standard that has done so much to popularize home theater. Dolby's own DPLIIx FAQ states in part: "Dolby strongly believes in the intrinsic value and benefit of the 5.1 surround platform.... The introduction of 7.1 playback by Dolby should in no manner be interpreted as an abandonment of the 5.1 playback medium, or of its obsolescence as a format." DPLIIx is simply an adaptation to existing market trends. But don't let the hyping of new surround formats goad you into tearing up a perfectly good 5.1-channel system. Any existing investment in 5.1-channel surround should provide pleasure for years to come.

The newest member of the Pro Logic family is **Dolby Pro Logic IIz**. It does everything IIx does and adds two **height channels** above the main front left and right channels. DPLIIz, explains Dolby, "processes low-level, uncorrelated information—such as ambience and some amorphous effects like rain or wind—and directs it to the front height speakers." This makes the front of the soundfield a little airier though the effect is usually subtle—I spent my first day with DPLIIz rushing to the left height speaker and putting my ear against it. With the right movie, it can be occasionally transformative, but not with all movies. With music, the effect is benign but not a game changer. DPLIIz is not encoded into movies at present though it is encoded into a few games. However, Dolby Atmos, the latest Dolby surround technology, does encode real spatial information into height channels—see the Dolby Atmos section below. The Dolby Atmos upmixer is replacing the Pro Logic family in some surround gear. More on that later.

The DTS counterpart of Dolby Pro Logic II and IIx is **DTS Neo:6**. It has a slightly "wetter" sound than DPLII, with more reverb in the surround channels, and less closely approximates the original feel of stereo mixes than Dolby Pro Logic II.

DTS Neo:X goes beyond Dolby Pro Logic IIz by deriving both height *and* width channels from stereo or surround sources. Neo:X converts source signals with 2.0. 5.1, 6.1, or 7.1 channels to 9.1 or 11.1 channels. In 11.1 channels, it creates what DTS calls "a semi-spherical soundfield." However, 11.1-channel receivers are rare. If you buy a 9.1-channel model, you'd have to eliminate one of the following: height, width, or back-surround. With a 7.1-channel receiver, eliminate two.

103

Audyssey's counterpart to Dolby Pro Logic IIz and DTS Neo:X is **Audyssey DSX**. It adds height and width channels. More on this later.

Dolby Digital 5.1 and DTS 5.1

These two surround standards dominate DVD and other digital program sources. They have 5.1 channels. Their 6.1- channel extensions will be discussed in the next section.

As originally designed by Dolby Laboratories, Dolby Digital is a 5.1-channel surround format that delivers three **discrete** (meaning separate) channels of information to the front left, center, and right speakers and two discrete channels of information to the surround left and right speakers. Dolby Digital also provides a discrete subwoofer channel labeled **LFE** (**low frequency effects**). That makes it an improvement over previous Dolby surround formats which used **matrixing** to derive the front-center, surround, and sub channels from front left and right.

To efficiently provide all this audio information, Dolby Digital uses **perceptual coding** to eliminate audio data that are masked by louder sounds and therefore unheard by the human ear. This so-called **lossy** encoding is not an audio meat grinder—it is based on sophisticated psychoacoustic principles and can provide surround sound of great subtlety and power. Dolby Digital usually provides 5.1 or 6.1 channels at 384-448 kilobits per second, quite a feat compared to, say, the MP3 format, which needs at least 192 kbps just to make stereo sound adequate. The standard's full range is 96-640 kbps. Any Dolby Digital-compatible surround receiver can accept up to the top limit. DVD is limited to 384-448 kbps, though Blu-ray's top limit is 640 kbps.

Dolby Digital performs other kinds of signal processing. Some, such as **dialogue normalization**, work automatically in the background. Dialogue normalization makes adjustments to overall levels to smooth out differences between program sources, keeping the words of performers intelligible, among other benefits. **Dynamic range compression**, also known as **midnight movie mode**, is switchable and under the viewer's control. This function reduces the disparity between the louder and softer portions of the soundtrack, using volume-level codes embedded in Dolby Digital soundtracks by the mixing engineer, while ensuring that the soundtrack retains some of its original dynamic range.

Next we move from Dolby Digital 5.1 to **DTS 5.1**, also known as **DTS Digital Surround** or just plain **DTS**. For nearly everything Dolby offers in the consumer sphere, there's a competing flavor of DTS, and it's important to understand the rival standards in context.

DTS 5.1 appears as an alternate soundtrack on some DVD-Video releases, in addition to Dolby Digital, and occasionally it pops up as an alternate soundtrack on DVD-Audio releases, instead of Dolby Digital. Most DTS soundtracks have 5.1 channels. (We'll get to the 6.1-channel versions of DTS and Dolby Digital in the next section.)

DTS purports to provide a higher-quality method of surround encoding. Like everything from Dolby Digital to MP3, DTS uses perceptual coding to eliminate audio data that are inaudible to human ears—in other words, it is classified as a lossy format. DTS is usually encoded at a higher **data rate** (the number of bits per second) than Dolby Digital. The range is 768 kilobits per second to 1.5 megabits per second, versus a maximum of 640 kbps (and in practice, more often 384-448 kbps) for Dolby Digital. Some say this produces less edgy sound, but that also makes it less efficient, reducing the number of bits that can be allocated to video on a DVD, and thus reducing picture quality. Both codecs have variable data rates, allowing mastering engineers to decide how much space on a DVD is devoted to sound and how much to picture.

Whether DTS-encoded DVDs are superior to those with mainstream Dolby Digital is hotly debated by home theater buffs, not to mention the Dolby Labs and DTS people. It would be fair to say that Dolby Digital is better entrenched in DVD movie releases as well as broadcast, cable, satellite television, and streaming.

Dolby Digital EX and DTS-ES

These surround formats add extra channels and speakers to the standard 5.1-channel array. Referred to as **6.1-** or (incorrectly) **7.1-channel** formats, they originated in movie theaters where there's a genuine problem in achieving even surround coverage for the folks in the back rows. However, is your home theater the size of a real moviehouse? Does it seat hundreds? Didn't think so.

Dolby Digital EX was developed by Dolby Labs and THX. Originally licensed through THX as **THX Surround EX**, it is now also licensed through Dolby as Dolby Digital EX, while THX has made EX a feature of its revamped **THX Ultra2** and **Select 2** certifications.

In any event, EX adds one extra channel, and one or two extra speakers, in what has become known as the **back-surround** position. While the regular surround (or in this context, **side-surround**) speakers are placed *toward* the back of the room, along the side walls, the back-

105

surround speakers go against the back wall. The two back-surround speakers receive the same signal. Technically, that makes Dolby Digital EX a 6.1-channel system—if you count data channels, not amp channels or speakers—though it is often referred to as a 7.1-channel system.

Recognizing that not every home theater system can accommodate an extra pair of back-surround speakers, some surround processors include a **phantom back center** mode that accepts the EX signal and simulates back-surround effects using the two side-surround speakers. This is a viable option to tearing up an existing 5.1-channel system.

In any event, EX derives that back-surround signal from the surround left and right channels using the same **matrixing** process that Dolby Pro Logic uses to derive front-center and surround signals from the front left and right. Doing away with matrixing is one of the key improvements that Dolby Digital has made over Dolby Pro Logic. Thus, in attempting to take a step forward, EX takes a step backward, emphasizing the quantity of surround channels over quality.

The 6.1-channel version of DTS is **DTS-ES Discrete**. It encodes the back-surround as a fully separate and independent channel. That makes it superior to EX which uses matrixing to derive the sixth channel from the surround left and right channels. For compatibility with non-discrete DTS-ES programming, DTS also provides a matrixed version of the back-surround channel (**DTS-ES Matrix**).

Dolby Digital Plus, Dolby TrueHD

Now we move into the newest generation of surround standards.

Dolby Digital Plus first found favor in European satellite delivery. In the United States, it made its first impact in the dueling high-def disc standards, the surviving Blu-ray and now deceased HD DVD. It is also approved for use in ATSC 1.0, the standard for over-the-air DTV broadcasting in the U.S., as a robust channel (with error correction). Netflix and Vudu use it for a/v streaming.

Dolby Digital Plus is not so much a replacement of the prevalent Dolby Digital standard as an extension. It supports up to 14 channels. The data rate can run from 96 kilobits per second to 6 megabits per second, and because it is more efficient, it works in satellite video, streaming media, and other low-bandwidth applications.

But doing the same with less is only part of Dolby Digital Plus's attraction. It can also do more with more, improving sound quality by increasing the data rate from Dolby Digital's average of 384-448 kbps to as much as 6 Mbps. That huge increase in data should provide a larger, warmer soundfield with more vocal clarity while virtually eliminating any audible effects of compression.

Dolby Digital Plus may help Dolby Labs leapfrog its competitor DTS by supporting full discrete channel separation in 7.1, 6.1, or 5.1 surround formats, something DTS-ES Discrete can do (in 6.1) but Dolby Digital EX cannot do (in 6.1 or 7.1). Mixes are totally independent—so if you prefer 5.1 channels, you'll get a dedicated 5.1-channel mix, with all channels in balance, not 7.1 stripped down to 5.1. DD Plus is also better at handling high frequencies, transient elements, and continuous tones than regular Dolby Digital. To maintain compatibility with existing Dolby Digital gear in tens of millions of homes, DD Plus converts to DD at 640 kbps through optical and coaxial outputs. While that may be a downconversion, it's higher than the previous 448 kbps maximum of Dolby Digital, so if the source is 640 kbps or better, the difference should be audible—at those data rates, every additional bit helps.

All of the digital surround encoding formats discussed above are lossy formats. That is, they use compression (or more precisely, data reduction and psychoacoustic coding) to send lots of data through narrow channels. As digital pipelines grow—most significantly, with the Blu-ray disc format—it becomes possible to use lossless encoding to deliver more data with no sacrifice in fidelity. What you hear is a bit-for-bit reconstruction of the original master signal with *nothing* lost. That is the promise of **Dolby TrueHD** and its competitor DTS-HD Master Audio (the latter is discussed in the next section).

Dolby TrueHD, like Dolby Digital Plus (above), is a fully discrete 5.1-, 6.1- or 7.1-channel format on Blu-ray though theoretically it can support up to 14 full-range channels (or 34 in Dolby Atmos). It musters data rates up to 18 megabits per second—more than 40 times the data rate of Dolby Digital—for what Dolby calls "studio master" quality. In addition to sheer size, the TrueHD bitstream is also smarter. It uses metadata, or data about data, for dialogue normalization (to even out volume levels among sources) and dynamic range (to adjust the range of soft-to-loud).

A key potential use of Dolby TrueHD in Blu-ray is to replace **uncompressed PCM**. PCM is the common language of many digital audio formats. The CD, for example, is a PCM format with strings of 16 zeroes and ones repeated 44,100 times per second. That makes CD a 16-bit/44.1kHz format. Richer PCM audio as high as 24/96 or 24/192 is being encoded in Blu-ray releases without compression, but this is a vast waste of space. TrueHD delivers the same quality in 25-50 percent as many bits. That leaves more space on the disc for better picture quality or extra features with no sacrifice in audio performance.

Blu-ray mandates the original Dolby Digital at an improved 640 kbps, but makes TrueHD (18 Mbps) and Dolby Digital Plus (1.7 Mbps) optional, which means the player may or may not have TrueHD/Plus decoders. In practice, most now do.

Recent surround receivers have onboard DD Plus and TrueHD decoders. The player passes the undecoded bitstream directly, instead of a converted high-res PCM or analog signal, to the receiver through the **HDMI 1.3** (and up) interfaces. With the decoder in the player, an HDMI 1.1 or 1.2 connection will provide full performance with Dolby Digital Plus and TrueHD. If the player has high-quality digital-to-analog converters, then an analog connection will also provide high performance, as long as your player or receiver has bass management to handle the crossover between satellite speakers and subwoofer. Using old-style optical or coaxial digital outputs will deliver to the receiver a downconversion to Dolby Digital at 640 kbps, still higher than DD's original 384-448 kbps in standard-def DVD, with your receiver's usual bass management.

Support for Dolby TrueHD and Dolby Digital Plus in Blu-ray players was initially spotty. However, they're universal in current players. Dolby TrueHD has taken on added importance as the vehicle for **Dolby Atmos**, the next generation of surround technology. Existing Blu-ray players can support Atmos with no need for upgrades.

DTS-HD High Resolution Audio, DTS-HD Master Audio

Let's climb the DTS family tree: We've already discussed **DTS 5.1** and **DTS-ES** (6.1) above. They operate at 768 kbps to 1.5 Mbps. Their sampling rate and bit depth, unmentioned above, are 48 kHz and 24 bits. There is also a **DTS 96/24**, unmentioned above, used as an alternate soundtrack on some DVD-Audio high-resolution music discs that doubles the sampling rate.

You may see programming labeled **DTS Encore**, **DTS Encore ES**, or **DTS Encore 96/24**. They are identical to the three listed above. DTS uses the Encore designation to distinguish its older formats on program packages. Historical footnote: The DTS codec has an alternate name, **Coherent Acoustics**, but it's never used to label hardware or software. That brings us up to date on the old DTS and its nomenclature, old and new.

These older (but still relevant) standards use what the DTS people call **core data**. But there are two newer standards that add **extension data** to vastly increase performance. The core and extension data are encoded and delivered in a single bitstream. Older surround hardware decodes only the core data. Newer surround hardware can decode the extension data, accessing two new DTS standards.

One of these, roughly comparable to Dolby Digital Plus, is **DTS-HD High Resolution Audio**. It encodes up to 7.1 channels at 96/24 resolution and data rates up to 6 Mbps on Blu-ray. And it is lossy, which means it uses compression and perceptual coding, just like abovementioned members of the DTS family.

But the ultimate DTS is lossless, reducing data but not omitting a single bit of the master source. This is **DTS-HD Master Audio**. It encodes up to 5.1 channels at sampling rates of 192 kHz and up to 7.1 channels at 96 kHz and data rates of up to 24.5 Mbps on Blu-ray.

The bitrates for these new members of the DTS family are variable. This allows them to kick in more bits when the soundtrack becomes more challenging—with an explosion, perhaps, or the roar of a crowd. The formats share a constant bitrate of 1.5 Mbps.

How high you're allowed to climb on the DTS family tree, qualitatively, is a matter of hardware, software, and connections. Except for DTS 5.1, all the DTS standards are optional in Blu-ray. However, DTS has been fairly explicit about connectivity. For DTS-HD Master Audio you'll need the latest HDMI 1.3 (or higher) interface in both player and receiver. DTS-HD High Resolution Audio travels at up to 6 Mbps via HDMI 1.1 and 1.2 (and higher). Software labeled DTS Encore (DTS 5.1 and DTS-ES) can travel by HDMI 1.1 and 1.2 or through standard old-style optical and coaxial digital jacks. And any of them operates, at full resolution, through analog jacks.

Hardware and software notes: As with the next-gen Dolby formats, DTS-HD may not be fully supported in all Blu-ray players. Some output DTS-HD MA as lower-res DTS core, even after firmware upgrades, which is very disappointing. But the latest generations of product have caught up, finally providing the full potential of DTS-HD.

And that's a good thing because DTS-HD Master Audio is now the dominant lossless surround codec in the Blu-ray format, with studio support far surpassing Dolby TrueHD. It sounds fantastic, as good as the source material will allow.

The latest in Dolby surround technology: Dolby Atmos

Dolby Atmos is the latest in cinema sound from one of the great powers in surround. It made its moviehouse debut in 2012 with Pixar's *Brave* and came to home surround equipment in 2014.

Until now home surround sound has focused on channel-based speaker configurations: 5.1, 7.1, etc. But this, Dolby says, "has gone as far as it can in the home." The future of surround lies in an **object-oriented surround** standard that uses metadata to specify the location and movement of objects in the soundfield without assigning them to any particular channel. The object audio renderer in the decoding and playback system then optimally adapts and scales these objects as well as possible with whatever speaker configuration is being used. One of the advantages of Dolby Atmos is that it is "one mix to rule them all." Instead of mixing separately for 5.1, 7.1, and Atmos, engineers create just one Atmos mix, which then works for everything.

Atmos is good for more than just action-movie sledgehammer effects. Dolby suggests it might also enable the listener to follow the flight of a hummingbird. It is not just about technology—it's about storytelling. Here's how Dolby's Brent Crockett put it in an interview with *Sound & Vision*: "Should this sound come from the left rear-surrounds or the left side-surrounds? With Dolby Atmos, filmmakers just have to think about the story: Where is that yelling child going to run? How will the helicopter move overhead after takeoff?"

In theaters, Atmos supports up to 128 channels and 64 speaker feeds, stacking ceiling speakers atop the main ones to create a height layer. This liberates surround effects from the boundaries of the room. Instead of operating solely in the horizontal plane, they are panned through a three-dimensional soundfield in, around, and above the audience. It's an all-enveloping sensation quite unlike ordinary surround.

In home gear, of course, most people will not add dozens of speakers. Atmos speakers are added in pairs. If you're starting with a 5.1-channel array, adapting to Atmos usually means adding one or two pairs of height channels and speakers. A minimal Atmos configuration is **5.1.2 channels** (three front speakers, two surrounds, one subwoofer, two height speakers). This can be powered by a seven-channel receiver, the most common kind. However, Dolby recommends a minimum of **5.1.4 channels** (three front speakers, two surrounds, one subwoofer, four height speakers). This requires a nine-channel receiver. Atmos in the home can support up to 24 speakers on the floor and 10 speakers overhead (**24.1.10 channels**)—though this would be for an elaborate and costly dedicated home theater installation.

Height speakers may be built into, or attached to, the ceiling. For ceiling mounts, choose speakers with wide dispersion patterns. Note that ceiling speakers may not work well in low-ceilinged rooms because the speakers would be distractingly close to listeners; in that situation, try Atmos-enabled speakers instead.

If ceiling mounting is not convenient, some speaker makers are accommodating height channels by building upward-firing drivers into the

Dolby Atmos, the latest innovation in home surround technology, can work with any surround speaker configuration but requires height speakers for the full 3D effect. Shown here is a 7.1.4-channel configuration combining seven floor speakers, plus sub, with four ceiling-mounted height speakers at front and back. This would require a receiver with 11 amp channels. Dolby's recommended minimum for Atmos is 5.1.4, omitting back-surrounds, requiring a nine-channel receiver. A 5.1.2-channel configuration, with only two height speakers, can get by on a seven-channel receiver. Many other configurations are possible because Atmos is not a channel-based technology.

tops of speakers. These **Dolby Atmos-enabled speakers** bounce sound off the ceiling at an angle that reaches the seating position while adding nothing to the speaker footprint. Atmos-enabled speakers work best with flat (not vaulted or angled) ceilings at heights of 8 to 9 feet, though up to 14 feet may be OK. Light fixtures, moldings, and vents do not interfere. Top-firing drivers may be built into existing speakers (**Atmos integrated speakers**) or added to them (**Atmos add-ons**).

While its channel and speaker configurations are flexible, any Atmos-compatible system requires a surround receiver or preamp-processor with Atmos decoding plus enough amp channels to drive the desired number of speakers. The downside of adding more channels is that, especially in receivers, they're all running off the same power supply, reducing the system's overall dynamic capability. 5.1.2 may not be much of a stretch—it's just reshuffled 7.1—but going beyond that may stress the receiver, even if it nominally has enough channels. So if you want more speakers, make sure you have enough power to run them. That points toward a separate preamp-processor and power amp, or a pre-pro and a rack of stereo amps. Speakers with high sensitivity or efficiency ratings would also help. While Atmos needn't be limited to a dedicated theater installation, its possibilities can be more thoroughly explored in a system that supports lots of high-quality amplification.

Delivery vehicles for Dolby Atmos are the same as for other Dolby surround technologies. On Blu-ray, it would be encoded in Dolby TrueHD; for streaming, in Dolby Digital Plus. Dolby says existing Blu-ray players should be able to carry Atmos signals, with certain extensions, as long as the players fully conform to Blu-ray specs—but be sure to turn off the secondary audio mode. Initial Atmos releases on Blu-ray arrived in 2014 from Paramount and Warner with dozens more coming in subsequent years. The first Blu-ray title with Atmos was *Transformers: Age of Extinction*. Vudu offers Atmos video streaming.

A Dolby Atmos soundtrack is fully compatible for playback in existing surround systems—it won't disable your system—though to get the height effects, you'll need Atmos processing, extra amp channels, and height speakers.

If you're playing non-Atmos channel-based content on an Atmos system, the **upmixer** deploys it to all speakers if desired. The activation of the height channels adds a touch of air to the top end. The effect is subtle—it probably wouldn't offend anyone. Dolby has resurrected the **Dolby Surround** name (which originally referred to analog surround technology at the birth of surround) to label this feature in surround gear. It either replaces, or supplements, the Dolby Pro Logic II family.

Most receiver manufacturers have announced Atmos-compatible models or firmware upgrades. Note that seven-channel receivers are limited to the Atmos 5.1.2 configuration; the full 5.1.4 configuration requires nine amp channels. Speakers are arriving from Atlantic Technology, Definitive Technology, GoldenEar, KEF, Klipsch, Pioneer, Teufel, and Triad. Onkyo offers an Atmos home theater in a box system.

The latest in DTS surround technology: DTS:X

The DTS version of object-oriented surround technology is **DTS:X**.

Whereas Dolby has very specific speaker layouts in mind for Atmos, DTS:X is more adaptable to whatever speaker layouts are expedient in the home environment. That includes Atmos layouts. But it's OK if, say, one surround speaker is higher or lower than the sweet spot. You'll be able to place speakers for best dialogue intelligibility.

DTS:X uses the encoding of soundtrack elements as separate objects to make some of them user-adjustable. If you're into action movies, and you find that loud effects are overwhelming voices, it's easy to boost dialogue. Dynamic range is also separately adjustable (most receivers also include a different form of dynamic range control). Additional user controls may be unveiled in the future to take advantage of other encoded objects.

DTS:X started rolling out to movie theaters in mid-2015. The first DTS:X-ready surround receivers became available around the same time, though a software update was required. By the time you read this, both DTS:X-ready and full DTS:X receivers should be available from several brands. At presstime not much was known about DTS:X speakers—however, given its flexibility in speaker layout, DTS:X should work well with Dolby Atmos speaker configurations. DTS:X programming is starting to become available on Blu-ray, traveling via DTS-HD Master Audio, and is backward compatible with existing DTS bitstreams and speaker layouts. The first title was *Ex Machina*. Roughly a dozen titles were available at presstime with more to come.

Auro-3D: An alternative to Dolby Atmos and DTS:X

Object-oriented surround sound is two Goliaths and a David, the latter being **Auro-3D**, from Auro Technologies of Belgium. Auro is from the people who operate Galaxy Studios, selling recording and production gear in Europe.

The Auro-3D technology was born in 2006 but is just starting to make a modest yet interesting impact on home theater. What does this 9.1- to 11.1-channel format offer that the spiffy new Dolby Atmos and DTS:X standards lack? Jaws dropped at the 2015 Consumer Electronics Show when an Auro demonstration added not just a single height layer, but two, combining with the floor speakers to produce a three-level height experience. It was among the most convincing demos of any surround technology I've ever heard.

The first Auro-3D products include the Auriga surround receiver ($16,700), which includes Dolby Atmos decoding; and the Mensa and Crux surround processors, which sell for five-figure sums in Europe. They include an Auro-Matic mode to upmix non-Auro material to the speaker configuration. The format runs on Blu-ray players, which do not require an update, with the data carried as generic PCM audio on Blu-ray discs. A hardware format without software titles is meaningless, but Auro's website listed about 40 movie and music titles at presstime. It also listed more than 100 movie titles mixed in Auro-3D, though it was unclear how many of them would make it onto Blu-ray.

Will this acorn grow into a might oak? Only time will tell, but it is a very promising technology that might find a bleeding-edge niche. For more information see auro-3d.com.

Dolby Volume

The dual purposes of **Dolby Volume** are to smooth out volume levels among sources, and to tame the dynamic extremes of movie soundtracks. The reference level used to calibrate surround systems is too loud for many listeners. But when you listen at lower volumes, your ears can't hear bass and treble frequencies as well. Dolby Volume adjusts these frequencies dynamically while keeping background sounds stable, avoiding "pumping" effects. Dolby is licensing it for surround receivers and other products. Competing technologies include **THX Loudness Plus**, **Audyssey Dynamic Volume/EQ**, and **SRS TruVolume**.

DVD-Audio and SACD

In addition to Dolby TrueHD and DTS-HD, there are two music-only surround formats that transcend the mundane. DVD-Audio and

SACD were invented to provide a higher quality medium for audio-only use. Machines handling both formats (plus DVD-Video and sometimes Blu-ray) are called **universal players**. For compatibility, your surround receiver or preamp-processor will need HDMI (1.2 or above). As an alternative, you can have the player output a high-res PCM signal to the receiver. Most receivers lack DSD decoding for SACD but the high-res PCM workaround provides about the same quality. Still another alternative is a set of 5.1-channel analog line inputs to mesh with the player's 5.1-channel analog line outputs. (Some older DVD players, receivers, and surround processors from Denon, Meridian, and Pioneer use their own proprietary 5.1-channel digital interfaces.) See DVD-Audio/SACD section in "Picture & Sound Sources/Disc players."

THX certification

THX started out as a program to correct inadequacies in film playback, ensuring that surround sound in theaters meets the standards set by filmmakers. Then it expanded into home theater with a product-certification program that makes it possible to implement Dolby Digital, DTS, and other surround formats in a more uniform manner. The home THX program, like its cinematic counterpart, ensures that home theater audiences hear what the filmmaker intended.

Dolby's and DTS's licensing requirements are designed to allow their surround formats to fit in a variety of products. THX's licensing requirements are designed to fit Dolby or DTS technology into a high-end home theater system. The aim of THX is to allow the home theater audience to hear what the engineer heard on the mixing stage, with plenty of high-impact bass and enveloping surround sound.

THX expanded from its original mission of product certification to the development and licensing of surround encoding formats. THX was a co-developer, with Dolby Labs, of the seven-speaker Surround EX format and took the lead in licensing that technology. (EX is now licensed by both THX and Dolby.)

Even as the cinematic THX program continues to cover commercial moviehouses and mixing facilities, the home THX certification program now covers everything from speakers to receivers, pre-pros, power amps, Blu-ray and DVD players, DVRs, and things like speaker cables, interconnect cables, and DVD movie releases. There are THX-approved DVRs, PCs, soundcards, and multimedia products—plus TVs and projectors from Epson, JVC, LG, Panasonic, Runco, and Sharp. There is even a new certification program for Ultra HD displays.

THX stands for Tomlinson Holman's eXperiment, after the audio scientist who originated it while working for Lucasfilm. It's also a reference to *THX 1138*, the first film of George Lucas, whose Lucasfilm company originally controlled the development and licensing of THX.

THX comes in several forms. **THX Multimedia** is for desktop use, with the screen two feet, four inches away. **THX Speaker Bar** certification is for one-piece horizontal speaker systems that accompany flat-panel TVs. **THX I/S** is for receiver and speaker bundles, with the screen six to eight feet away. **THX Select** is the kind you're most likely to see in speakers and DVD players, while **THX Select2** is for receivers. This version of THX is designed to work best in smaller rooms, up to 2000 cubic feet, with the screen 10 to 12 feet away. The original THX spec is now called **THX Ultra**, and found in surround preamp-processors as well as high-end receivers and other products. THX Ultra is suitable for larger rooms, up to 3000 cubic feet, with the screen more than 12 feet away, and is different in other subtle ways. It has been phased out in favor of **THX Ultra2,** a 7.1-speaker extension of Ultra that is suitable for multichannel music as well as movies. Products licensed to include THX Loudness Plus are certified as **THX Select2 Plus** or **THX Ultra2 Plus**. All varieties of THX require certified gear to play at high volume levels, to disperse sound in specific ways, to maintain a low level of noise and distortion, and to behave in a stable and predictable way.

Here are some of the major aspects of THX certification:

- re-equalization
- timbre matching
- rear-surround decorrelation
- bass management
- amplifier specs
- reference level
- THX Optimizer/Optimode

Re-equalization compensates for the differing tonal balances that are appropriate in home theaters versus moviehouses. It can reduce the "brightness" of many film soundtracks. In a moviehouse, the soundtrack has to have a somewhat brighter treble to enable the audience hear all the dialogue. But in a home theater, which doesn't have the hundreds of bodies and other absorptive elements of a moviehouse, that same

116

soundtrack may seem too forward, abrasive, and oppressive. Re-EQ is so useful that it's become available even in non-THX gear in both officially licensed and unlicensed variants.

Timbre matching attempts to match the tonal qualities of the front and rear speakers to create a more uniform surround soundfield.

Decorrelation of the surround channels is a clever way of introducing a little diversity to the single mono surround channel that serves the two rear speakers in Dolby Pro Logic surround. Why bother? Because a mono signal precisely centered between two surround speakers can collapse the channel (and thus the part of the soundfield) closest to the listener. The effect of decorrelation is to make the soundfield in the back of the room more diffuse, which is a good thing. Newer THX products have **adaptive decorrelation**, which recognizes when the circuit is not needed—in other words, when fully discrete surround channels come into play.

Bass management is (in part) a generic term for the crossover settings that a surround processor uses to route bass from the front and surround speakers to the subwoofer. THX recommends using a subwoofer and setting all other speakers to "small" during system setup. This allows you to place the sub where it sounds best in your room. When larger full-range speakers (bookshelf or bigger) are used with a sub, the sub is usually shut down. One of the great benefits of THX is that the spec is very consistent about bass management, in terms of both the 80 Hz crossover point (the frequency at which the sub takes over from the main speakers) and the slope (the way one driver's response declines as another driver takes over). In some products, bass-management options go beyond the basic crossover to include additional subtleties. For instance, you might have the option of mixing bass information from the front left and right channels with the sub channel.

Amplifier specs are tougher in THX gear than in non-THX gear. A THX-certified receiver or power amp must be able to drive a speaker with an impedance of 3.2 Ohms (that's a very demanding load) and produce a volume level of 105 dB (that's very loud). That will handle just about any movie-soundtrack or musical peak known to humankind. THX Select2 equipment must achieve this feat in a room of 2000 cubic feet (height times width times depth). THX Ultra2 equipment must do it in a room of 3000 cubic feet. The room-size numbers are a guideline, not a dictate, but with THX it helps to know how big your room is.

The THX **reference level**—synonymous with the Dolby reference level—is intended to replicate the level at which the movie was mixed.

117

Using it will keep dialogue at a moderate level, while effects and music may have more dynamic punch. Both Dolby and THX specify a reference level of 85 dB (decibels). At the **controller** (THX-speak for preamp-processor), the volume meter registers the number of decibels above or below reference level (plus or minus one, etc.). When setting the THX reference level in home theater, be warned that the processor may include a 10 dB offset to spare the ears. Then you would set 75 dB on your SPL meter to obtain the 85 dB reference level.

THX Optimizer, or **Optimode** is for DVD movie releases. It places test patterns on the disc to enable the home theater buff to tweak the video display to the same settings used by the mastering engineer. This is a brilliant and useful idea.

THX Ultra2 certification

Here are some of the major aspects of THX Ultra2 certification in addition to the basic THX requirements listed above:

- EX auto detection and switching
- high-bandwidth video switching
- adaptive side-surround decorrelation
- boundary gain compensation

EX auto detection and switching detects a flag in EX-compatible disc releases and switches on the EX back surrounds. When the flag is missing, the user may switch on the EX circuit manually if desired.

High-bandwidth video switching requires a device to pass high-definition and/or progressive-scan video signals without degrading them. THX does not specify the necessary bandwidth but does test licensed products to ensure that no degradation is visible.

Adaptive side-surround decorrelation switches itself on when the surround processor receives mono surround signals (for example, in Dolby Surround-encoded software). It affects only the side-surrounds, not the back-surrounds. When the processor receives stereo surround signals (for example, in Dolby Digital 5.1-encoded software), they are sent to both the side-surround and back-surround channels, and decorrelation is switched off. This assumes that the movie has stereo surrounds and that EX is switched on.

Boundary gain compensation is a switchable circuit that tames excessive bass. Applied before the processor's bass management circuit, it is intended for listeners sitting near the rear wall where acoustic conditions tend to result in exaggerated bass response. Many rooms are not large enough to allow placement of a sofa in the middle of the room. If your sofa is placed near the rear wall, boundary compensation will provide more accurate bass for listeners sitting there.

The latest from THX

THX has always been oriented primarily to audio. But in 2006, that began to change with the company's first certifications of video products, flat screens, projectors, and Blu-ray players.

As with any kind of THX certification, manufacturers submit products in the prototype stage for testing. In video products, THX testers are on the lookout for motion artifacts, ghosting, dropped frames, and noise. Their front-of-screen testing covers brightness, contrast, color gamut (range), color gamma (balance), uniformity, and resolution. Video signal processing tests include scaling, deinterlacing, and conversion, especially as it affects motion.

Another strategic shift for THX, starting in 2007, is a deeper involvement in product design. THX already maintains labs to certify products. Manufacturers are turning to THX for help with mechanical, electronic, and firmware design.

THX Loudness Plus, which bowed in 2007, recognizes that the reference level is too loud for many listeners. It compensates for tonal and spatial shifts that occur when you listen at lower volumes. It is available in THX Ultra2 Plus and Select2 Plus certified products. Other licensed versions of competing technology include **Dolby Volume** and **Audyssey Dynamic Volume/EQ**.

In 2010 THX announced an initiative aimed at the nascent 3DTV category. In conjunction with BluFocus, THX offers three new categories of certification. **THX-BluFocus AV Certification** ensures that sound and image measure up to the master source. **THX-BluFocus Creative Certification** ensures that 3D elements not deviate from the director's intent or cause viewer fatigue. Finally, **THX-BluFocus Interoperability** certifies that discs play on both 3D and 2D players.

THX Media Director uses metadata—or data about data, similar to the way your iPod keeps track of artist and song names—to adjust aspect ratio, volume reference level, picture settings, and surround modes. The program has been in the works for several years. In 2011 it passed a couple of milestones: Chip makers began building the processors necessary to get Media Director into 3DTVs, surround receivers, and Blu-ray players. And the Blu-ray release of *Star Wars: The Complete Saga* was the first software to embrace this useful standard.

Finally, in 2011 THX added a new certification category. **THX Speaker Bar** takes aim at the soundbars being sold to complement flat-panel TVs. Certified products must have the ability to cover a wide listening area, flat (or at least flatter) frequency response, strong dialogue clarity, high output with minimal compression and limiting, proper blending at the bar/subwoofer crossover, and a lower crossover to prevent voices and other elements from shifting location from bar to sub.

To learn more about the THX technology portfolio, visit thx.com.

Audyssey's auto-setup and listening modes

In addition to Dolby, DTS, and THX, there is another player in surround technology. Audyssey was founded by R. Tomlinson Holman and Chris Kyriakakis of USC—Holman had previously cofounded THX. Audyssey licenses auto setup, room correction, and low-volume listening modes as well as other listening modes that add width and height enhancements to surround's basic 5.1-channel configuration.

Audyssey got its start by licensing combined auto-setup and room-correction technologies to makers of surround receivers. These make it easy to set up a receiver by measuring the speakers and room, then making adjustments for speaker size, speaker distance, and room acoustics. Versions start with **Audyssey 2EQ**, which takes measurements from up to three positions in the listening room, and provides "basic resolution filters" for satellite speakers but not for the sub—all other versions filter both sats and sub. **Audyssey MultEQ** measures from up to six positions (or 32 if you're a custom installer) with "mid-level resolution filters." **MultEQ XT** measures from up to eight positions (32 for CI), with "high-resolution filters." The latest version, **Audyssey MultEQ XT32**, makes the best of the limited audio processing powers in receivers, using higher sampling rates where they will do the most good, and providing 32 times as much filter resolution as its predecessors. Denon's names for these technologies are **Audyssey Bronze**, **Audyssey Silver**, **Audyssey Gold**, and **Audyssey Platinum**. Audyssey is the class act of its field—if you want auto setup and room EQ, it's a reliable choice.

Two other Audyssey modes take aim at bass-related problems. **Audyssey Sub EQ HT** facilitates adding a second subwoofer—among the best ways to improve any system. It tailors level and delay and is in-

120

tegrated into MultEQ. **Audyssey LFC** prevents bass frequencies from assaulting neighbors through your walls. Dynamically monitoring bass output, it targets and adjusts problematic frequencies while also using psychoacoustic processing to restore in-room perception of low bass.

When a system is running at reference level, the dynamic extremes of movie soundtracks can be fatiguing. You turn the volume down to avoid the loudest effects, then turn it up again to catch dialogue, and this quickly becomes tiring. Audyssey is among the companies licensing **low-volume listening modes**. **Audyssey Dynamic Volume** smooths out volume variations, while **Audyssey Dynamic EQ** adjusts frequency response and surround envelopment, so that soundtrack elements are not only audible but natural-sounding at low volumes. Turning on Dynamic Volume always activates Dynamic EQ; but when Dynamic Volume is off, Dynamic EQ can be turned on or off. So you can use Dynamic EQ without being bound by Dynamic Volume's level-setting decisions.

The latest version of Dynamic Volume is **Audyssey Dynamic Volume TV**. It does not affect varying volume levels between programming and ads, an issue already addressed by a new federal law (the **CALM Act**). Instead it smooths out volume fluctuations within programs. There is also an **Audyssey VXT** mode designed to help small speakers stand up to challenging loads. These two modes seem not to have been widely licensed at presstime. Other companies licensing low-volume listening modes include Dolby (with **Dolby Volume**) and THX (with **THX Loudness Plus**).

Audyssey has moved into surround post-processing modes with **Audyssey DSX**, the first to support both width and height channels. (DTS Neo:X also supports width and height; Dolby Pro Logic IIz also supports height.) What is the benefit of adding width channels? According to Audyssey, the ear can hear in more directions than current technology allows. "Experiments have shown that human localization is better in front than to the sides or behind. This means that for front-weighted content such as movies and most music, good engineering dictates that we employ more channels in the front hemisphere than the back. Imaging is also better horizontally than vertically and so good engineering also dictates that channels must first be added in the same plane as our ears before going to higher elevations." What Audyssey is saying is that, if you've got seven amp channels to work with, adding width will provide the greatest benefit, adding height is second best, and adding back-surrounds is third best. In my opinion, both width and

121

back-surrounds are dispensable, while height can be fun for some movies. None of them invalidates an existing 5.1 system—the incremental benefits are minimal. See the "Installation guide" for placement tips.

Minor proprietary surround modes

The following modes are found only in a handful of products. These are not mainstream surround formats that involve encoded software and decoding hardware, like the Dolby and DTS formats, though they may be fun for a few listeners.

The following technologies originated from SRS Labs but were acquired by DTS in 2012. **Circle Surround** adapts stereo sources to surround. The 5.1-channel version competes with DPLII while the 6.1-channel version (**CS II**) competes with DPLIIx and DTS Neo:6. Circle Surround has a brighter and more forward sound than the other modes, making it suitable for speakers that have a relaxed, laid-back sound. Then there's **TruSurround**, which enhances dialogue and bass; **TruVolume**, which smoothes out extreme volume fluctuations; and **WOW HD**, which expands the clarity and bass of stereo content.

Logic 7 derives 7.1 channels of sound from 5.1-channel or stereo sources. It is found in receivers by Harman Kardon and surround processors by Lexicon. **QuantumLogic Surround** is a new mode in the 2011 update of the Lexicon MC-12 preamp-processor. Starting with 7.1 channels, it adds three front height channels and two rear height channels, plus extra sub channels, for an eyebrow-raising total of 12.4.

Cinema DSP 3D is Yamaha's name for a mode that adds two front height channels to a 7.1-channel array. That means that the total of *front* channels rises from three to five. Yamaha refers to its regular DSP modes as Cinema DSP.

DSP modes

To complement the mainstream surround formats, most surround receivers and processors offer **DSP** (digital signal processing) or **EQ** (equalization) modes that mimic various surround ambiences such as concert halls, jazz clubs, etc., sometimes plus or minus a few channels. Some receivers have dozens of them. These are intended primarily for processing stereo music sources. The use of Dolby Pro Logic II is usually preferable because it's cleaner and less gimmicky.

Overdone DSP surround effects sound "wet"—dripping with reverb. They're entertaining for a few minutes but their crudity quickly becomes tiring. More subtle ones send only a whisper of ambience to the rear-surround speakers and are more listenable over the long haul.

DSP is also used to process mainstream surround formats. However, the phrase "DSP modes" usually refers to non-licensed additions.

Are five channels too many for you?

A system can have more than 5.1 channels or fewer. More than just exercises in reductionism, the less-than-5.1 modes are valid ways to build a practical surround system—adapting technology to your needs, not vice versa. (Note the two forms of nomenclature: mine first, Dolby Labs second. With Dolby's method, the sub is implied.)

Surround 5.0 (or 3/2) keeps the three front and two surround speakers, but eliminates the subwoofer, routing bass to the other speakers. Set all speakers to "large," sub "off." If two front speakers (or all five) are full-range tower or bookshelf models, why bother adding a sub, especially if you find action-movie dynamism more an irritation than an asset? The classic stereo pair can provide adequate to excellent bass. Using big speakers in lieu of a subwoofer is a far better option than buying a mediocre sub. But you will need either speakers with a high sensitivity/efficiency rating or a muscle amp to drive more demanding speakers.

Surround 4.0 (or 2/2), with two channels in front and two in the rear, was the original configuration of **Dolby Surround**, as it was then called. This was before Pro Logic added the front-center channel, and way before Dolby Digital. You don't have to dig up vintage Dolby gear to produce 4.0 surround. Just tell your surround processor that there's no front center speaker; or, if you're using an old Pro Logic processor, switch the front-center channel to **phantom** mode. In either case, this effectively shuts down the center speaker and redistributes its signal to the front left and right speakers. This might temporarily be a better alternative than leaving a center speaker atop a TV where a busy toddler might tug on its cable. Or you may not have an adequately matched center speaker. Or there may not be a position where a center speaker blends in acoustically—spaces can be unpredictable.

Surround 3.0 (or 3/0) uses the **Dolby 3 Stereo** mode found in some surround processors, though you can achieve the same effect by simply shutting down the surround pair but keeping the front left, center, and right speakers. Ditching the surrounds may be a necessity in some room layouts, especially those where speaker stands, invasive wall mounting, or long cable runs may not be practical (even a tiny screening room can easily require dozens of feet of cable to reach each rear speaker). Surround 3.0 provides all the benefits of stereo while eliminating the traditional "hole in the middle" weakness of two channels. If you have Dolby processing gear and three matching speakers, you can have a strongly continuous front soundstage. And to take advantage of it, there's lots of stereo on the air as well as on tape and disc.

Surround 2.0 (or 2/0) is stereo. There are still a few audiophiles (old and grey and very stubborn) who insist 2 channels work better than 5.1. Surround 2.0 goes beyond the embrace of plain-vanilla stereo to include enhanced stereo formats that simulate surround effects from 2 speakers. Examples include **Dolby Virtual Speaker**.

Picture & Sound Sources

*A*system is only as good as what it's fed. The mistaken belief that all digital sources perform the same is more entrenched in the home theater sphere than in the high-end audio world. We home theater buffs could well afford to take an attitude lesson from our audiophile friends who take pains to feed their systems the highest-quality signals available from turntables or high-end digital sources. Choosing the correct picture and sound sources also has ramifications in ease of use, as anyone who has ever struggled with a badly designed disc player knows all too well.

The highest-quality signal source is Ultra HD on Blu-ray, which is just getting started. UHD has also been available on hard drives provided by TV makers and in a lesser-quality form from streaming services. The next highest-quality sources of digital television signals are regular Blu-ray, which supports 1080p, and the airwaves, where HDTV in the 1080i and 720p formats is being broadcast. Most cable systems are now required to offer digital versions of local channels but UHD has been slow to emerge. DirecTV and the Dish Network offer packages of HDTV channels. The DirecTV version of HDTV is slightly truncated but still high-quality. HD-enabled, hard-drive-based digital video recorders are capable of capturing the full image resolution of HDTV (though regular VHS VCRs and some DVRs are not). While streaming offers some nominally Full HD and even UHD content, slow data rates may contribute to a crude picture. In the standard-def domain, progressive-scan DVD is the best choice, followed by conventional interlaced DVD and (if you're in a nostalgic mood) laserdisc. Standard-def video streaming lags behind and VHS is almost painful to watch.

Disc players

There are plenty of ways to get movies and music into a home theater system, but the primary hard-copy video formats in home theater are high-def **Blu-ray** and standard-def **DVD** (which stands for **Digital Video Disc** or **Digital Versatile Disc**, depending on whom you ask). DVD has so far been the fastest-growing format in the history of consumer electronics, exceeding both the VCR and the CD. Its high-def replacement, **Blu-ray**, is available in standalone players, PlayStation game consoles, compact in-a-box systems, PC drives, and a few HDTVs. It's also penetrating hitherto DVD-only territory such as jukeboxes and portables.

Ultra HD Blu-ray

The bad news about Ultra HD on Blu-ray is that it requires a new player—a software update won't do it. And of course you'll need software to go with the hardware. The good news is that both were becoming available as this was being written in late 2016.

The Blu-ray Disc Association imposes several requirements on UHD Blu-ray: Of course it must support **UHD resolution** of 2160 by 3840 pixels. But really great UHD requires a host of other picture-quality improvements too. They will be baked into the standard.

To improve brightness and shadow detail, a variety of **HDR video** standards are supported, both proprietary (Dolby Vision) and open (SMPTE 2084 and 2086, MaxFALL, MaxCLL). Color-related improvements include **color gamut** up to Rec. 2020, increasing the range of colors; and 10-bit **color depth**, increasing the range of shades within colors. HDR is required in players but optional on discs. Other requirements include **HEVC H.265** video compression, which makes it possible to efficiently transmit much of the above, and a 60 Hz frame rate for video productions (not including movies which are generally 24 Hz or 24 frames per second). To contain all this goodness, UHD Blu-ray supports both 66 GB dual-layer and 100 GB triple-layer discs (vs. the 25-50 GB of regular Blu-ray) though not a single-layer disc.

Important note: UHD requires **HDCP 2.2** digital rights management throughout your system, so make sure your UHDTV and surround receiver have it as well as your Blu-ray player. UHD requires the **HDMI 2.0** interface, and to take advantage of HDR video, make that **HDMI 2.0a**.

UHD Blu-ray mandates **backward compatibility** for 1080p Blu-ray. While DVD and CD compatibility are not mandated, manufacturers are unlikely to antagonize consumers by excluding them. 3D 1080p is an option but the UHD Blu-ray standard does not support 3D UHD.

An optional feature set called **Copy and Export** allows players to incorporate a hard drive, so they wouldn't be limited to playing discs. Panasonic and Samsung introduced models so equipped in Japan. That's the Copy part; the Export part allows consumers to use software services to obtain copies for play on portable devices. UltraViolet is one example of how this might work.

Ironically, many of the initial UHD Blu-ray releases are upconverted from 1080p masters (in other words, the source is 2K, not 4K). Another growing pain is the variability of HDR technology. "At its worst," wrote *Sound & Vision* video expert Tom Norton in early 2016, "it has me fishing around for the best settings on the display, which can vary more than I'd prefer from disc to disc."

With that excitement over, let's dig into the existing Blu-ray format.

Blu-ray: state-of-the-art high-def on disc

The high-definition videodisc was a long time coming. Unfortunately, it arrived in duplicate, as the consumer electronics industry shot itself in the foot yet again with yet another childish squabble over licensing revenue. Remember Beta vs. VHS? It was followed by Blu-ray vs. HD DVD, with Blu-ray emerging as the victor.

Samsung's UBD-K8500 ($400) is the first Blu-ray player to support UHDTV. It requires UHD Blu-ray discs to deliver full resolution.

127

The **Blu-ray** format uses a blue-light laser to store up to 25 GB on a single-layer disc, or up to 50 GB on a dual-layer disc. It supports Full HD 1080p video and the movie-friendly 24 fps frame rate as well as higher frame rates. To make the most efficient use of this bit bucket, it also uses video compression: either MPEG-2, MPEG-4 AVC, or VC-1 (see "Television/DTV by the numbers").

Initially Blu-ray had an Achilles heel. Because it read the disc at 0.1 mm—a much shallower depth than existing DVD, at 0.5 mm—the disc was fragile, and had to be enclosed in a cartridge, evoking sad memories of the deceased CED videodisc from the 1980s. However, TDK saved the day by developing a hard disc coating that eliminated the need for the cartridge.

On the audio side, Blu-ray mandates Dolby Digital at improved data rates up to 640 kbps, plus DTS and uncompressed PCM. Some discs include stereo soundtracks as well as surround. Optional audio codecs include the lossless codecs, Dolby TrueHD and DTS-HD Master Audio; and the improved lossy ones, Dolby Digital Plus and DTS-HD High Resolution Audio. DTS-HD MA has proven to be the choice of most studios, firmly establishing lossless surround in most Blu-ray disc releases. Players may also downmix surround soundtracks for output via stereo analog jacks.

Blu-ray's interactivity standard is **BD Java**. The original and now outdated version was Profile 1.0. Profile 1.1 supports **Bonus View** features such as expanded local storage, picture-in-picture, and internet access. It's also called Final Standard Profile, though it's not the final one. Profile 2.0, a.k.a. **BD-Live**, players support internet-enabled features such as chat and messaging. But the Blu-ray camp didn't designate it until October 31, 2007, and some manufacturers rush-released players to beat the deadline. So some older Blu-ray players won't handle BD-Live features on existing and future disc releases because the hardware lacks the requisite on-board memory and ethernet jacks. Profile 3.0 is for audio-only use.

For digital rights management, Blu-ray uses a scheme called **AACS** (**Advanced Access Content System**). Among many other things, this permits encryption that could cause an "attacked" player to shut down. AACS has been hacked several times. Blu-ray adds two more forms of protection. **BD+** can examine and verify player keys, or alter the player's output. It also has been hacked. The **BD-ROM Mark** is a cryptographic mark that is stored separately from other content and is needed to unlock playback. So far it remains unhacked.

AACS also includes a flag called the **image constraint token**—a ticking time bomb insert at the insistence of Hollywood. Players made after December 31, 2010 down-res high-def to standard-def through the analog component video outputs. Players made after December 31, 2013 eliminate analog-outs. This is called the **analog sunset**. If your older HDTV depends on component-out for HD, you may wish to buy an older player.

One of the coolest new Blu-ray and DVD features is **network connectivity**. This is your ticket to trendy video streaming from the likes of Netflix along with internet radio and a variety of other app-driven goodies. Some players with ethernet or wi-fi connections can access media from a PC through a router. **DLNA-certified players** can access music, video, and photos via the Windows Media Player.

Versions of Windows from Vista onward support Blu-ray. Mac does not, Steve Jobs having described the format as "a bag of hurt."

Alternatives to Blu-ray

Many DVD players upconvert their standard-definition output to high-definition formats. These are not true high-definition players but they may provide a smoother picture with some high-definition displays.

Even in the wake of HD DVD's demise, Blu-ray still has some faint competition. One is China's **CBHD**. It uses a blue laser to read a 30 GB disc and inherits some technology from HD DVD, thus allowing the developers to avoid playing licensing fees to the Blu-ray people. The audio/video codec is the homegrown Chinese **AVS**. Warner announced software support in 2009. CBHD is now outselling Blu-ray in China.

A few HD movie titles have been released in the **UMD (Universal Media Disc)** format developed by Sony for the PlayStation Portable gaming console. There have been no new releases since 2011.

There are also video compression formats for high-def recording on PC. The best known is **DivX**, from DivX Networks. It allows decrypted video to be re-encoded in the Windows Media Player, with **MPEG-4** video compression. The resulting efficiency makes it a viable format for file sharing. While players from at least a half-dozen major brands offer DivX playback, there is no prerecorded software in the format, so you'd be limited to PC-generated discs, presumably containing downloaded material—not all of it legal. DivX has been called the MP3 of video. For more information see divx.com.

Is there life after Blu-ray? Panasonic and Sony have developed a next-gen format, the **Archival Disc**, that would use multiple cartridge-enclosed discs to store 300 GB, though it is designed for professional, not consumer, use. Another possible replacement would be **holographic laser** technology. It uses multiple lasers to read and write data three-dimensionally, as opposed to addressing a single pit. And it promises to store far more data than any technology using a single red or blue laser. Again, its most likely use will be industrial, not consumer.

The DVD-Video format

The **DVD-Video** format (most just call it DVD) is a variation of the physical read-only disc format known as **DVD-ROM**. As originally designed, DVD-Video delivers **MPEG-2** compressed video of varying quality, combined with up to eight streams of sound usually encoded in **Dolby Digital AC-3**. (MPEG is a standards-setting outfit, the Moving Picture Experts Group; AC-3 is the original name of Dolby Digital.) However, DVD-Video sound is not limited to Dolby.

MPEG-2 is a form of **lossy** video compression used to fit movies within the storage capacity of a disc. When films are mastered for DVD release, the MPEG-2 encoder pares down the digital video information to eliminate redundant data such as backgrounds that don't move. When the encoding is done well, its effects are invisible. When it's done poorly, a variety of distracting digital **artifacts** result, especially in scenes containing rapid movement or other complex video information. Common artifacts include jagged diagonal lines (**jaggies**) and a picture dissolving into blocks. Video displays can introduce artifacts of their own.

DVD-Video is not an **HDTV** (high-definition television) format. It supports either of two officially sanctioned **DTV** (digital television) formats known as **EDTV** (enhanced-definition television) and **SDTV** (standard-definition television). Either way, DVD produces a picture with vertical resolution of 480 pixels and horizontal resolution of 720 pixels, roughly the limit of most conventional analog TVs, though any HDTV can do better—which is why Blu-ray exists.

Even within its standard-def limits, DVD's MPEG-2 video compression varies according to the skill of the mastering engineer, who allocates the disc's storage capacity. Scenes with rapid movement all over the screen take up more disc space than those with unmoving backgrounds. Another factor is the quality of the authoring software. If you thought one movie looked great while another looked awful, it's probably not your equipment that's at fault.

For **copyright protection**, the DVD format uses **CSS** (**Content Scrambling System**) which ensures that DVDs play back only on licensed DVD players. Also common is the **Macrovision APS** (Analog Protection System), which is designed to foil VCR-recording circuits, but also causes flashing and other unpleasant problems with certain video displays.

One of DVD-Video's more obscure areas of optional compatibility is the **VCD** (**Video CD**) format, which uses MPEG-1 video encoding with a fairly unsubtle form of video compression. Except as a file-sharing medium, VCD never really took off in the west. In the far east, however, it is wildly popular. Western home theater buffs will find that VCD's picture quality is not sufficient for big-screen movie viewing, but if you dare to explore the VCD format, you might find some interesting music videos and karaoke titles.

Audio features of the DVD-Video format

If the movie is a contemporary one, six channels (or 5.1, to use the usual nomenclature) will form a surround soundtrack. The other two may provide mono, stereo, or two channels encoded with Dolby Surround. Some discs include 5.1-channel Dolby Digital *and* DTS soundtracks (sometimes accompanied by a reduction in video quality as the **bit bucket**—the supply of digital data—runs low). Does DTS sound better than Dolby? Let the debate rage on; I'll have a Bombay Sapphire Gin martini, straight up, olive, please. DTS is standard equipment in audio/video receivers and DVD players. However, only a minority of DVD movie releases are encoded in DTS. Dual-5.1 discs may omit the two-channel alternate soundtrack. If no two-channel soundtrack is available for use with a stereo system, the DVD-Video player can mix a 5.1-channel soundtrack down to stereo. This is called the **Dolby Digital mixdown**. (None of this has anything to do with **DVD-Audio**, a separate 5.1-channel audio-only format discussed further down.)

While the DVD-Video format specs do not officially include **CD compatibility**, manufacturers provide it as a courtesy to the consumer. What about homemade **CD-Rs** and rewritable **CD-RWs**? The first few generations of DVD players handled them inconsistently, if at all. However, more DVD players have become compatible with CD-R and CD-RW discs. For extra assurance, either look for **MultiPlay** certification (by the Optical Storage Technology Association, osta.org). Or just take a few favorite brands of recorded CD-R to the store and try 'em.

Look carefully at your CD collection and you just might find dozens of **HDCD** releases. If that's the case, seek out an HDCD compatible DVD player or receiver. HDCD stands for **High Definition Compatible Digital**. Invented by Pacific Microsonics and now owned by Microsoft, it is a stealth format that shoehorns 20 bits worth of digital information onto a standard 16-bit CD. An HDCD sounds its best on an HDCD-compatible player but will play on any CD or DVD player. The improvement in sonics is real, if subtle, and has come with a refreshing absence of pain or hype—HDCD has not caused a format war, invalidated an older format, or raised the cost of CDs. And it's now embedded in more than 5000 CD releases. HDCD capability is a feature in some DVD and Blu-ray players and surround receivers.

Some DVD and Blu-ray players play music or picture files from burned discs. Check manuals for your favorite file formats.

More Blu-ray and DVD-Video features

Dual-layer Blu-ray discs can hold up to 50 GB while single-sided DVD-Video discs can hold up to 4.7 GB. That does not mean all discs hold that amount; capacity varies with the number of layers and sides. BD and DVD support both **dual layer** discs (the laser briefly delays the program to refocus from one to the other) and **dual-sided** discs (the user must flip the disc). A dual-sided disc sometimes may have a widescreen version of a movie on one side and a nonwidescreen version on the other. There are also dual-sided Blu-ray/DVD combo discs.

There's enough storage space on both BD and DVD to hold more than one version of a movie, soundtracks in different surround formats, foreign language soundtracks, captioning in multiple languages, up to nine camera angles, seamless branching of varying storylines, "making of" documentaries, and still photos. You might even find the hidden features that some buffs call **Easter eggs** (mostly trivia-related video clips). All players support these features—but disc releases vary. Use of the multi-angle feature, for instance, is still disappointingly rare.

Optical disc formats are easy and fast to navigate. Instant random access lets you flit from scene to scene—assuming **chapter stops** have been inserted at the most helpful moments by the mastering engineer. Some releases make a big deal of chapter stops, with illustrated menus.

BD and DVD both include **regional coding** which permits movie studios to designate different release versions for different parts of the world (North America, Europe, Japan, etc.). This may be convenient for marketing purposes, but it frustrates collectors, who seek out doctored players that defeat the regional coding feature.

The **parental lock** feature is intended to prevent children from viewing unsavory material. Parents must set up a passcode. Adult holders of the passcode can view anything they want—kids can't. Of course, passcode holders must keep turning the feature on and off if they don't want to live by the same standards they impose on their kids.

A more sophisticated parental control feature is program filtering. One such technology is from the Salt Lake City-based firm **ClearPlay**. A ClearPlay-compatible disc player can be set to use any of 14 levels of filtering to screen out violence, sex, or other material deemed offensive by some. The filters are compiled for every major new disc release by the company's staff of "movie professionals" and can be updated by a CD sent by U.S. Mail or by downloading them from a website and burning them onto a CD-R. Other products use **TVGuardian** filtering, which uses dialogue embedded in closed captioning to decide what language to mute. TVGuardian is available for Blu-ray as well as DVD.

Networking works through an ethernet wired connection, though some BD players also support wi-fi. This lets the player access media via internet streaming or from a PC using the DLNA standard.

PSSST! Your player may have a secret underground feature—the ability to skip ads and copyright warnings. Hit **stop, stop, play** at the beginning of a disc and see what happens. Whether it will work depends on how the disc was authored.

DVD-Audio and SACD

These two formats were invented to provide a higher-quality medium for audio-only music listening than the compact disc. Yes, we were promised CDs would provide "perfect sound forever"—and for the majority of listeners, they sound fine—but many audiophiles have been reluctant to give up their LPs, saying that CDs sound harsh, vague, grainy, or two-dimensional.

In addition to higher fidelity in general, DVD-Audio and SACD offer an opportunity to expand the two-channel palette of music recordings to as many as 5.1 channels. Both formats also support stereo recordings.

DVD-Audio and SACD are not the first formats to provide music in surround. In addition to a handful of LPs from the quad era, a handful of CDs have been released with Dolby Surround or DTS. However, only a few music producers have taken advantage of the opportunity to mix in those formats. To provide flexibility, the format specs for both DVD-Audio and DVD-Video allow the use of 5.1-channel Dolby Digital—but that audio-for-video format uses perceptual coding to eliminate supposedly inaudible audio data. DVD-Audio and SACD, in contrast, are music-delivery formats that do not eliminate any part of the original audio signal. While both (especially DVD-Audio) do support visuals, neither is intended for movie viewing.

DVD-Audio

DVD-Audio builds on the success of the DVD-Video format. DVD-Audio programming can be played only on a DVD-Audio compatible player (not all DVD or BD players qualify). The DVD-Audio format delivers anywhere from 2 to 5.1 channels of music encoded at far higher data rates than the current CD format. Whereas the CD delivers a string of 16 zeroes and ones 44,100 times per second, DVD-Audio delivers a string of 24 zeroes and ones 96,000 to 192,000 times per second—or three to six times more bits. Depending on the requirements of programming and disc capacity, DVD-Audio offers the option of using **Meridian Lossless Packing (MLP)** to fit more data onto the disc. Unlike the perceptual coding scheme used in Dolby Digital, MLP reconstructs the full digital audio signal, bit by bit, eliminating nothing.

The current DVD-Audio format was preceded by a stereo format advertised as a capability in many DVD-Video models. It was known under several names including **24/96** and **PCM**. (The latter is a generic name for a type of digital encoding also used by the CD format.) This earlier format's 24-bit encoding and 96 kHz sampling frequency beat the CD's 16 bits and 44.1 kHz hands down, offering subtle but audible benefits. However, the full-blown version of DVD-Audio does the same thing far more efficiently thanks to MLP, and can do it in 5.1-channel surround sound, not just stereo.

DVD-Audio releases often include both uncompressed DVD-Audio and one additional compressed soundtrack that may be either Dolby Digital or DTS. Discs that include an alternate soundtrack (as most do) allow an extra measure of flexibility. If you don't own a DVD-Audio player, you can still play the alternate soundtrack on your DVD-Video or Blu-ray player. Dolby Digital and DTS are both compressed formats, and therefore won't sound as good as DVD-Audio, but having an alternate soundtrack option allows you to stock up on DVD-Audio discs while you save your pennies for a DVD-Audio player.

One drawback of first-generation DVD-Audio gear was a lack of bass-management features—in other words, the ability to designate speakers as large or small and route bass to a subwoofer. In a system with full-range speakers, a powerful rhythm section might deliver an overwhelming quantity of bass. However, newer DVD-Audio gear does include bass management. Also—addressing another early criticism of the format—newer DVD-Audio players default to the DVD-Audio soundtrack, even when used without a video display.

While the major labels no longer support DVD-Audio, the format has seen an unexpected renaissance thanks to Porcupine Tree's Steven Wilson, who has authored fresh 5.1 mixes for classic albums by King Crimson, Yes, Emerson Lake & Palmer, Jethro Tull, and XTC. This has proven all over again that DVD-Audio is a great medium for music.

SACD

DVD-Audio has suffered from a format war with the **SACD (Super Audio Compact Disc)** format from Sony and Philips. SACD produces far more detail and subtlety than the CD thanks to its use of Sony's **DSD (Direct Stream Digital)** technology which records four times as much information as a regular CD. DSD originated as a reference-quality format for professional recording, mastering, and archiving. Frequency response extends to 100 kHz (five times the CD's 20 kHz) and the sampling rate is an astronomical 2.8 million bits per second. (A direct comparison with DVD-Audio is rather tricky due to DVD-Audio's multiple sampling rates.)

SACD has some so-called **multichannel management** features that are reminiscent of mainstream surround formats in general. These include **channel-level adjustment** (from the listening position), **channel-balance adjustment** (front-to-back and side-to-side), and **bass management** (which can redirect bass from main speakers to sub). The format supports graphics and video but does not require a video display to be running when SACDs are playing.

SACDs can have dual layers to double running time. But the coolest thing about SACD is that it can (and sometimes does) come on a **hybrid disc** with a regular CD layer and a souped-up SACD layer. Thus it can play on any CD player, even in a car-audio system or a boombox. If you find titles that interest you on hybrid SACDs, you can stock up on them, play them on your existing CD equipment, and get the full benefit of SACD when you buy a compatible player.

Both DVD-Audio and SACD can benefit stereo or surround-encoded music recordings. Neither has been embraced with much enthusiasm by the music industry, which is more concerned with getting control of file sharing. But there continue to be trickles of releases, especially on SACD, from European classical labels. You can even add both formats to your system in one box with **universal disc players** that handle DVD-Audio and SACD as well as DVD-Video (and more).

SACD-compatible disc players can output signals in either SACD's native DSD or in high-res PCM. Only a few surround receivers accept DSD but just about all do high-res PCM, allowing users to experience the pleasure of high-res audio.

The DSD file format is not limited to SACDs. It is also used for a modest number of high-res music downloads and is supported by a few high-end component USB DACs (digital-to-analog converters for computers). This is an area of future growth for DSD.

DVD Plus vs. the DualDisc

It's possible to combine CD audio and DVD features (either DVD-Video or DVD-Audio) on opposite sides of a single disc. The beauty of it is that you can slip it into any CD player to enjoy traditional (CD audio, yet when it hits your DVD player, it can produce video and multichannel audio. But there's a catch and it's that same old sad refrain—this hybrid CD/DVD was killed by yet another asinine industry format war.

The first format to emerge in the marketplace was **DVD Plus**. This bonded disc can hold 78 minutes of audio on its CD side and 135 minutes of audio and video on its DVD side. And then there's the **Dual-Disc**. This is another bonded disc, though it holds only 60 minutes on the CD side, and a single layer (about 120 minutes) on the DVD side. It was test-marketed in in February 2004 and a small flurry of releases followed in 2005. Disc releases have ceased in both formats.

Progressive-scan DVD-Video players

Owners of digital television sets should check out **progressive-scan DVD players**, a significant high-end alternative in DVD-Video. These players are able to recognize, and compensate for, the various forms of mayhem that happen to movies as they go from film elements, to master videotape, to the MPEG-2 video compression format at the root of DVD-Video, en route to your DTV screen. Specifically, high-end players deliver pictures to the video display in full frames. In doing so, progressive-scan DVD players provide a more "filmlike" image.

Older DVD-Video players deliver the picture in an interlaced scanning format, known as **480i** (SDTV), intended to suit now-gone analog TVs. Progressive-scan players, which now dominate the market, **de-interlace** the picture and deliver it to the display in a **480p** (EDTV) format that better suits digital TVs. (Some go a step further, upconverting to 1080p for 1080p sets.) In addition to delivering a full-frame progressive picture to the video display, progressive-scan players also eliminate any lingering effects of the interlacing and 3/2 pulldown processes that may have crept in prior to or during disc mastering. Therefore a really good progressive-scan player has to perform some fancy footwork to deliver a clean progressive picture to your screen.

The next logical questions are: What is interlacing? What is the 3/2 pulldown? And why is it so important to eliminate all traces of them?

Interlacing is the old analog process that chops up each frame into a pair of separate half-frames called **fields**, with gaps between the scan lines. The video display reunites each pair of fields into a full **frame**, with no gaps between scan lines. Efficiency in use of over-the-air spectrum is the reason why interlacing was invented. That's why interlaced scanning is a staple of analog television as well as certain DTV formats. But the interlacing process is an undesirable element in digital video because it can distort moving objects and camera moves. These interlacing-related video distortions are known as **motion artifacts**. On a big screen, their presence can become painfully obvious.

The **3/2 pulldown** is necessary to convert film (which runs at 24 frames per second) to video (which runs at 30 frames, or 60 fields, per second). It stretches film's 24 frames across analog video's 60 fields by repeating one field in every other frame. A pair of matching fields alternates with a trio of mixed fields, hence the name 3/2 pulldown.

Alternating normal frames with a jumble of patched-together fields can have messy results. It makes smooth pans look jerky and still frames look flickery. And this frame-rate conversion is unnecessary if you're watching a properly mastered DVD (with 24 frames per second) on a DTV (which converts everything to one of its own native formats).

In DVD mastering at its best, the 3/2 pulldown is reversed before the film content reaches the MPEG-2 encoder that compresses the video signal for disc mastering. The result should be 24 frames per second on the disc, similar to the original film content, though the frames may be encoded as 48 interlaced fields, arranged in pairs. Each pair represents a moment in time, and each field is marked with a progressive flag that enables a full frame to be reconstructed. Some progressive-scan players use the flags to recognize the paired fields. More sophisticated players can recognize fields without using the flags. In either case, prog-scan players convert the image to full-frame progressive scanning at 60 frames per second.

In the best of all possible worlds, progressive-scan DVD players would not have to undo the effects of interlacing and the 3/2 pulldown because they wouldn't make it onto the disc. In the real world, these analog quirks live on in expensive telecine and videotape mastering equipment. On top of that, stuff just happens—errors can enter at any stage of the mastering process as the movie goes from film elements to tape master to disc master. A good progressive-scan DVD player has to be nimble enough to detect errors and compensate for them.

So here are a few things that might happen to your favorite movie when you watch it through a progressive-scan DVD player: It starts as film, running at 24 frames per second. The film is transferred to videotape using a scanner, which may run at 24 frames per second or 60 fields per second. And the videotape recorder that receives the signal may run at 60 fields per second. Then, if the MPEG-2 encoder receives 24 frames, it encodes them as 24 frames. If it receives 60 fields, they are converted back to 24 frames. If any errors occur during these processes—and they do—then the progressive-scan DVD player will analyze the motley succession of fields and convert it back into full frames … more or less. The video display performs the final act of conversion, upconverting the progressive-scan signal to its own native scanning rate.

This is an awesome (and often unpredictable) series of digital video processing events. How the progressive-scan player handles the third phase of it—recognizing and reversing any aftereffects of interlacing and the 3/2 pulldown—is where things get really interesting.

The following is a necessary oversimplification, but in ascending order of quality, a progressive-scan player may: Simply repeat each scanning line of a single field to form a full frame (**line doubling**). Generate a new line, dot by dot, looking at both the previous and next lines, and making an educated guess (**interpolation**). Or generate a new line by looking at individual **pixels** (dotted picture elements) across three or more fields (**field-adaptive de-interlacing**). The latest de-interlacing circuits go a step further, gulping up two whole frames worth of video

138

data to predict how moving objects will change from frame to frame. Slicker ones are good at reproducing mixed film and video content (such as pictures with superimposed lettering) without blurring either part.

Today's best consumer-level progressive-scan players provide a picture with greater vertical and temporal resolution, less flicker, and a nearly invisible line structure that encourages the viewer to sit closer to the screen. The video image has more of the look and feel of film.

Finally, remember that while a good progressive-scan DVD player can improve upon a DTV with a bad line doubler, some DTVs have better de-interlacing chips than some DVD players, and therefore may actually look better when fed with the player's interlaced output.

The many faces of DVD and BD for PC

In addition to the **DVD-ROM** (read-only) drives that have spread throughout the PC world, there are several **recordable DVD** and **recordable BD** formats. In order of seniority, they are:

- DVD-R (1997)
- DVD-RAM (1998)
- DVD-RW (2001)
- DVD+RW (2001)
- DVD-R DL, DVD+R DL (2004)
- BD-R (2005)
- BD-RE (2002)
- BD-R LTH (2008)
- BDXL (2010)
- IH-BD (2010)

The **DVD-R** format came first. Its specification was designed by the DVD Forum, a consortium of more than 200 electronics companies. DVD-R and its rewritable companion **DVD-RW** are often designated as the "dash" formats. Then Sony, Philips, and Hewlett Packard came along with the "plus" format, **DVD+RW**. Needless to say the dash and plus formats are mutually incompatible. But there are black-box recorders as well as PC drives that handle both, so if you play your cards right, you needn't worry about betting on the loser. **DVD-RAM** has random access capability, allowing it to rewrite portions of a disc, and making it superior for computer data recording. It has made inroads into Panasonic video recorders. The **DVD-R DL** (Sony's name) and **DVD+R DL**

(Philips' name) are dual-layer formats that hold up to 8.5 GB, enough for four hours of DVD-type (MPEG-2) video. With list pricing dropping well below $300, the DVD recorder has taken over from the VCR as a recording format with removable storage.

Blu-ray has recordable variations, not all of which available to consumers. **BD-R** is a write-once disc while **BD-RE** can be rewritten many times. Disc capacity is 25 GB for a single-layer disc, 50 GB for dual layer, 100 GB for triple layer, and (in BD-R only) 128 GB for quadruple layer. The **BD-R LTH** write-once disc uses an organic dye layer. Older Blu-ray gear may not support it though the Sony PlayStation3 is a notable exception. The newer **BDXL** is either a 100-128 GB write-once disc or a 100 GB rewritable disc. **IH-BD** (**Intra-Hybrid Blu-ray**) combines write-once and rewritable layers, both 25 GB.

Hackers have made it their business and their passion to break the copy-prevention feature that allowed DVD to make its debut amid a brief honeymoon between consumer electronics manufacturers and Hollywood movie studios. The first well known Hollywood-vs.-hackers case involved a copy-prevention-cracking utility called **DeCSS** (CSS stands for Content Scrambling System). As the case was winding its way through the courts, word came that the same decrypting can now be done in seven lines of code.

Connections

- HDMI output
- DVI output
- IEEE 1394-DTCP output
- ethernet/wi-fi network connection
- component video output (progressive or interlaced)
- S-video output
- composite video output
- digital outputs (optical or coaxial)
- stereo analog audio line outputs
- 5.1-channel analog audio line outputs

The **HDMI** digital audio/video interface is standard equipment in high-def Blu-ray players and standard-def DVD players. The most desirable versions of HDMI are 1.3 and up if you want DTS-HD Master Audio, 1.4a for 3D, or 2.0a for UHD/HDR, though any version will feed video to an HDMI-compliant TV if both devices allow copy prevention.

DVI is rare. The **IEEE 1394** interface, also rare, allows for networking. The ethernet (and on some players, wi-fi) connections allow software updates. Some players also use their network connections to pull content off a router-connected PC's hard drive or stream programming from Netflix, et al.

High-end players also provide **progressive-scan output** through their HDMI or **component video** jacks. BD players made after 2010 do not support HD through component video due to Hollywood skulduggery. They may also have **S-video** and **composite video** outputs for analog TVs, but those disappeared after 2013—more skulduggery.

All Blu-ray and DVD players include **digital outputs** that feed a receiver with Dolby Digital and DTS surround and CD stereo signals. These outputs may be the **optical** or **coaxial** types or both. Note that these do not convey lossless surround at full resolution. Many add extra digital outputs to feed a digital recording device or a digital-to-analog converter. Unfortunately, the digital outs of both DVD-Video and DVD-Audio players deliver a truncated 20-bit/48kHz version of the 24-bit/96kHz signal to allay the music industry's copyright concerns. That's above CD-quality but below the maximum audio quality level of which DVD-Audio is capable. (Occasionally, with some players and discs, if copyright-protection flags are not in place on the disc, the player will pass a true 24/96 signal. That's the exception, not the rule.)

The **stereo analog audio** outputs usually provide a two-channel mixdown of surround formats, as well as CD signals converted to analog. The two-channel mixdown may include embedded Dolby Surround, compatible with Dolby Pro Logic processors.

In Blu-ray, the **5.1-channel analog audio line outputs** are valuable for outputting lossless surround codecs like Dolby TrueHD, DTS-HD Master Audio, and uncompressed PCM to pre-HDMI receivers with 5.1 analog-ins. The surround-capable analog jacks may also be useful in other situations. For example, DTS-HD MA only travels through HDMI 1.3 and up, so if your player and/or receiver lacks HDMI 1.3 and high-res PCM, you'll need the 5.1-channel analog interface to get the high-res signal from player to receiver. Not all Blu-ray players have 5.1 analog-outs, so if you need them, study spec sheets.

HDMI 1.2 will pass SACD signals, and HDMI 1.1 will pass DVD-Audio. Players can also be set to output these formats as high-res PCM, which is compatible with most current receivers. In players without HDMI, a set of analog outs is necessary to feed high-resolution surround signals from a DVD-Audio and/or SACD player to the 5.1-

channel inputs on a surround receiver (whether the signal is surround or stereo). A select few universal disc players and receivers support the use of a single digital connection—for example, Pioneer, and Denon Link—but this is regrettably rare. The analog line connections can also be used for Dolby Digital and DTS movie soundtracks. However, if analog line connections bypass the receiver's crossover and bass-management settings, you'll be limited to the disc player's bass adjustments.

Shopping

The most sophisticated shopper will want to put players through a short but revealing series of test scenes. (An old favorite is the opening scene from *Star Trek: Insurrection,* where the camera pans across an urban landscape whose geometric forms easily reveal jagged digital artifacts that result from unsubtle video processing.) In addition to BDs and DVDs, try CDs, as well as anything else you want to play, such as the various recordable DVD and CD formats or MP3s burned onto disc.

How quickly does the machine boot up, load a disc, and respond to disc transport commands (play, scan, etc.)? That will affect the experience of using it. If you're not comfortable with its speed and rhythm, you may not enjoy using the player.

Front panel controls have become minimal on today's slimline disc players, so the **remote control** is now crucial. Does it fit your hand, differentiating transport controls by size and shape, placing these and other commonly used functions where your fingers expect to find them? In darkened home theaters, backlit or glow-in-the-dark keys can help.

DVRs, streamers, & servers

Recording devices add convenience to a home theater system. Originally that meant adding a VCR. Today it means adding a DVR, server, or streaming device. DVRs and streaming devices offer ways of temporarily capturing and displaying content. Servers, on the other hand, exist to store and organize content, including disc collections, and play it.

DVRs

A **DVR** (**digital video recorder** or **PVR** (**personal video recorder**) is a tapeless system that goes well beyond the capabilities of a VCR. If your phone rings, a DVR can pause the action, leaving a still frame while you're occupied. When you return, the DVR can pick up where you left off, or continue the program in real time, or go back to the beginning while still recording the remainder. While there are DVD-based and VHS-based digital recorders, the terms DVR and PVR usually refer to hard-drive-based components.

Recording capacity is a function of how big the unit's hard drive is. Cable systems are just starting to implement **RS-DVRs**, or **remote-storage DVRs**, which replace local hard drives with server storage. Hollywood and the networks tied up this cost-saving idea in the courts but it finally broke free.

DVRs use video compression, as do digital video formats and satellite systems, but there's a tradeoff between the **quantity and quality** of recording—as running time goes up, picture quality goes down. A bad DVR picture looks different than a bad analog VCR picture; rather than going fuzzy, it pixellates, dissolving into an expanse of blocks.

Recording with a DVR is easy thanks to the **onscreen program guide**. Read through it, click on the items you want, and the machine records your selections. There's no need to fuss with start/stop times or type numeric codes, as in the bad old VCR days.

The TiVo Bolt, starting at $199, is a state-of-the-art DVR. It supports UHD, has an ad-killing SkipMode feature, and even bids to replace your cable box.

Most DVRs now record **high-definition TV** formats though some channels may be only standard-definition TV—with about the same resolution as a DVD, or less, depending on what recording-time/quality option you choose. DVRs from cable, satellite, and telco providers also support video on demand.

There are a lot of ways to get a DVR into your system. Hard drives are cheap, so a handful of technically literate home theater buffs prefer to build their own PC-based DVRs with generic parts and ATSC tuners.

However, most DVR fanatics prefer an easier route. If you are a satellite subscriber, check out satellite receivers with integrated DVRs. Likewise, if you're addicted to cable, ask your cable operator about the cost of a DVR-equipped cable box. You can also buy component DVRs. If you want your DVR to have a more sophisticated and helpful user interface, check out TiVo.

TiVo: the deluxe DVR

The first generation of DVRs included three branded standalone products that were not limited to a single cable or satellite service: **TiVo**, **ReplayTV**, and **UltimateTV**. The latter two have since retreated under new ownerships, leaving TiVo as the class act of the field. UltimateTV is now owned by Microsoft and marketed as an extra-cost option through DirecTV. ReplayTV left the self-branded hardware realm, evolved into PC software, and is now owned by DirecTV. It program guide has been discontinued.

TiVo still sells standalone DVRs but also has arrangements with some cable and satellite (DirecTV) operators. For the standalone version, TiVo charges a **subscription** fees for its program guide. If you want to avoid subscription fees, several manufacturers offer DVRs with a **PSIP**-based program guide—that stands for Program and System Information Protocol and it's a data signal transmitted by broadcasters on their on DTV channels. A few DVRs come with built-in DVD players or recorders.

Basic TiVo functionality starts with the program guide, which serves for both program discovery and recording. You can search by title, keyword, performer, etc. You can preview upcoming programs and schedule recordings during previews. TiVo streamlines the process by recording favorite series, skipping repeats, and resolving recording conflicts. It is smart, learning your viewing preferences and making suggestions. Recorded shows are indexed alphabetically or chronologically.

TiVo allows online/mobile scheduling of recordings and transfer to mobile devices.

The latest version of TiVo is the sixth-generation Bolt. It can send recordings not only to home TVs but remotely to smartphones and tablets. It supports streaming from Netflix and other major services. New features include unified search and watchlist, ad-killing SkipMode, and recording up to four shows at once in UHD. Storage is 1 TB, or 150 hours of HD, and you can add an external hard or NAS drive.

Connections for TiVo and other DVRs

- HDMI outputs
- CableCARD slot
- RF antenna inputs
- component, composite video outputs
- optical digital line outputs
- stereo analog line outputs
- RJ-11 phone jack
- IR blaster
- ethernet, wi-fi network connections
- USB port (TiVo)

Some DVRs have CableCARD slots to function with local systems while others may have RF-ins for antenna or satellite. Broadband models require a high-speed network connection. Most DVRs come with high-quality HDMI and analog component video outputs for HD. Composite video jacks are usually provided. Audio outs can be optical or analog. The RJ-11 phone jack allows the system to update its electronic program guide. The IR blaster controls your cable or satellite box. The USB and ethernet ports added to newer TiVo products allow access to digital photo and music files through a home network.

Movie streaming and download services

DVRs grab programming from over the air, satellite, or cable. Streamers and servers get it through a broadband internet connection. Some services let you **stream** (to rent) movies or TV episodes while others let you **download** (to own) them.

There is some overlap between streaming and downloading services. For instance, Apple's iTunes Store and Amazon offer both streaming and download of movies. Amazon offers *The Wizard of Oz* in HD and SD for $2.99-3.99, streaming; or $9.99-12.99, downloading. Content streamed or downloaded from Amazon can be viewed on various TVs and set-top boxes. Content streamed or downloaded from iTunes can play on various Apple devices such as the iPhone and iPad.

Retail chains that once just rented movies on disc now allow them to be streamed. The biggest player in this field is **Netflix**. If you'd prefer a la carte pricing, consider **Blockbuster On Demand** or **Vudu**. **Hulu** with ads is free; **Hulu Plus** with fewer ads requires a small monthly fee.

Another major player in streaming is **Roxio CinemaNow**, which operates in PCs, TVs, set-top boxes, Blu-ray players, smart phones, and other media devices. Pricing is a la carte. Best Buy has licensed Roxio's technology, allowing consumers who buy the chain's Insignia-brand TVs and Blu-ray players to set up the CinemaNow service at the store. Covering all bases, Insignia products also stream Netflix.

If you prefer downloading to streaming, you'll need a way to store your downloads, and perhaps a way to make them available throughout the home. Do this with a **network attached storage (NAS) drive**, a type of external hard drive that acts as a file server on a home network.

Cable and satellite companies offer movie streaming through their **video on demand** services. Subscribers who pay an additional fee per program can access it through a set-top box or authorized DVR.

The line between streaming and downloading may blur as the entertainment industry explores **cloud-based content services**. Examples include the **Amazon Cloud Drive/Player** and the **Apple iCloud**. You pay a fee to purchase rights to enjoy a piece of content, accessing it from a variety of broadband-connected devices such as TVs, Blu-ray players, PCs, gaming consoles, smartphones, and tablet computers. You can also upload content you already own.

Another major cloud-based initiative is **UltraViolet**, a "buy once, play anywhere" licensing system supported by several major movie studios, TV makers, cable systems, retailers, and others. Buy a BD or DVD with UV and you can access the content via cloud on other platforms. UV is supported by most studios with the notable exception of Disney which developed its own **Disney Movies Anywhere**. UV's Common File Format now supports DTS-HD surround sound.

Music streaming services

The largest source of free audio streaming is **internet radio**, which can be accessed via PC, server, a/v receiver, or standalone radios. Radio Denmark alone has three classical channels. There are also services that aggregate internet radio stations with a convenient user interface. They include **vTuner** and **SHOUTcast**. For a more personalized internet radio experience, try **Pandora** or **Slacker**.

If you don't mind paying, there are also **music subscription services** that operate online. You can access them via PC, mobile devices, or some models of broadband-connected surround receivers and Blu-ray players. **Spotify** is currently the ranking favorite, especially in its ad-supported free version. For higher-quality lossless streaming, the audiophile favorite is **Tidal,** though it is limited to CD-quality 16-bit audio.

Satellite radio services from **Sirius** and **XM** have merged along with the two companies. They offer a cornucopia of music and talk via subscription. The internet version can be added or bought separately.

Music download services

File sharing started the downloadable-music phenomenon but paid downloads now provide a meaningful amount of revenue to the music industry. The most successful example is **iTunes**, Apple's music retailing arm. **Amazon** and others offer MP3 downloads, which are DRM-free and therefore not device-dependent.

Be warned that some of these services impose restrictions on the use of downloaded material. Most notably, some downloads will work only with some devices. The **iPod** and iTunes are the most notorious example, using a form of the **AAC** file format protected by Apple's **Fairplay DRM** (digital rights management). iTunes now offers **iTunes Plus** downloads which are both DRM-free and encoded at a higher bitrate, 256 kbps, for higher audio quality. Older purchases can be converted to iTunes Plus, though at a price, and not on a track-by-track basis—you have to convert the whole library.

Other services used Microsoft's **WMA** (Windows Media Audio) with **PlaysForSure** DRM, though most of them have either closed or switched to no-DRM MP3. The hazards of DRM have become apparent as some services (Microsoft, Yahoo Music) have gone out of business— and their encryption keys expired, leaving legal purchasers with tracks that couldn't be played or moved to new devices.

Don't neglect the musician-run sites. A lot of up-and-coming bands offer free tunes, and if you want to support the band, you can always buy a T-shirt. **YouTube** and social networking sites have become significant sources of music for budget-conscious listeners.

High-res music download services

The next step in the evolution of online music retailing is the new wave of high-res download services such as HDtracks. They typically offer either lossless or uncompressed music files. This chance to improve what we listen to comes as more people are shifting from component audio systems to computers and mobile devices.

Until now the demand for higher-quality downloadable audio has been limited by storage capacity. But hard drives and flash memory have gotten cheaper, so you can store more and/or higher-quality audio files. Thus it now makes more sense to buy high-res FLAC or other files.

Improvements in computer storage, audio software, and audio hardware are a virtuous circle. That is, they are mutually reinforcing. When you've got more storage space, you've got more room for high-res audio files. When you're paying big bucks for FLACs, you want them to sound great. So you might consider adding a USB DAC, starting at just a few hundred bucks, to transform your computer audio for the better. Or you might get into high-end headphones. And once you've done that, you might look into high-end headphone amps. And now that your DAC and 'phones are really rockin', you'll want to buy more FLACs.

Be warned that the term high-res refers to a variety of music file formats and other characteristics. Some have higher resolution than others. I define high-res as a lossless or uncompressed file with a bit depth of 24 (and up) and a sampling rate of 48 kHz (and up). See the "Introduction to audio file formats" and other sections a few pages below.

Introduction to servers and streaming devices

Multiple devices vie to become your digital signal source of choice. You can stream a/v or audio-only services from a smart TV or Blu-ray player, as discussed in chapters on those products. But you can also use a component media or audio server, set top box, game console, phone, tablet, or computer. Each of these platforms has different ramifications for content, quality, and ease of use. For example, if you want to stream the Netflix-funded series *House of Cards*, only Netflix has it. If you want HD quality, some platforms purport to have it, but the devil is in the specs (and the compression). And it's easier to stream a movie from a wireless mobile device than from an HDMI-connected computer—yet an increasing number of audiophiles prefer to stream music from a computer with an external USB DAC to improve audio quality. You will gravitate toward the options that serve you best.

Digital media servers

The term "server" usually refers to computer at the heart of a network. That's why the term is crossing over into home theater. **Digital media servers** handle both video and audio—**Kaleidescape** is an outstanding example—though copyright issues surrounding the copying of video from disc to server have kept them tied up in court. There are also digital audio servers (see next section).

The advent of Ultra HDTV brought a largely unfulfilled need for UHD signal sources (at least until UHD Blu-ray arrived). To fill the gap, Sony offered a UHD media player with two-terabyte hard drive. Samsung gave UHDTV buyers a USB hard drive preloaded with programs.

Digital audio servers

Audio files can enter a home theater system in several ways. A digital audio server is the deluxe method. It not only gets files into your system by ripping discs, but helps you organize your library too. An easier and more casual alternative is to stream from a computer or smartphone through wireless Apple **AirPlay** or **Bluetooth** connectivity. **Capacity** dictates how much material you can load into a storage device. Recent products tip over into multiple-terabyte territory. Digital audio servers sometimes contain a **CD-R/-RW drive**. It can be used either to load music into the system (from CD to hard drive) or to burn discs (from hard drive to CD-R or CD-RW). A few models dispense with the built-in CD drive but are able to connect to (and control) CD jukeboxes containing hundreds of CDs. Products with **upgradable firmware** can be updated to provide compatibility with new file formats.

Digital audio streaming players

There has been a recent shift from digital audio servers, which store and serve music, to digital audio streaming players, which access music from other devices such as computers and NAS (network attached storage) drives. This allows flexibility in what you use to **store** your music—you're not limited to the server, and you can upgrade capacity or change devices on a whim. It also allows **wireless connectivity**, a must for the modern listener. And it may even come with an attractive **interface** that makes it easy to access your music, using either a dedicated **touchscreen** or tablet/phone **app**. A growing number of audio streaming players come from high-end audio companies as audiophiles wean themselves off (or at least supplement) old-school LP and CD gear.

Tablets and smartphones as media and music servers

When you've got a tablet or phone in your hands all day, using it to supply your entertainment fix at night comes naturally.

The popularity of iPads and iPhones as a/v source components didn't happen by accident. Apple has steadily built bridges to home theater systems in several ways. First came the iPod/iPhone docks. Then surround receivers began to build iOS compatibility directly into their USB jacks. AirPlay Express blazed a wireless path and AirPlay itself became a built-in receiver feature. The Apple TV set top box—which Steve Jobs bashfully described as a "hobby"—has further extended AirPlay's reach. Once iOS showed how well a touchscreen interface could deliver entertainment, consumer habits began to change. iPhones and iPads even support Bluetooth, a non-Apple technology, making those devices irresistible wireless double threats.

For Android tablet and phone users (version 4.2 or later), there's **Miracast**, now starting to appear on smart TVs and other devices. Unlike Apple TV, it is not brand-specific, so devices from more than one manufacturer can work together. It uses a device-to-device connection based on the Wi-Fi Direct standard, and duplicates the content of a tablet or phone screen on the TV, which is called **screencasting**. Both 1080p video and surround sound are supported. Some products have Miracast baked in while others accommodate it in the form of an app. It is included in Windows 8.1 and 10. Android devices can also use the **MHL** wired connection, a variation of HDMI, to feed a TV or receiver.

Most cable and satellite operators offer apps that extend paid-for programming to a subscriber's tablet, smartphone, or computer browser. This can be a great way to take your video fix on the road. Ask your local operator what devices and programming are supported.

Set top boxes and dongles as media and music servers

Set top boxes serve up a wide variety of audio/video programming.

With Boxee having been acquired by Samsung, presumably to be cannibalized for smart TVs and Blu-ray players, the two leaders are Roku and Apple TV.

Roku makes the most popular streaming set top box. It started with audio streaming devices and worked its way up to the current Roku audio/video set top boxes, which support SD, 720p HD, 1080p HD, and now UHD. Roku is also available in compact form as the Roku Stick. Roku's dizzying array of programming includes dozens of free and premium channels. For more information see roku.com.

Apple led the way with the **Apple TV** box which streams movies, music, and photos. The **AirPlay** wireless system underlying it requires a home network connection, either ethernet or wi-fi. In addition to media from a computer running iTunes or iCloud (of course) Apple TV also supports content from Netflix, Hulu Plus, YouTube, and others. The original version had a hard drive which has since been replaced by flash memory. The new version offers video support up to 1080p plus Dolby Digital 5.1 surround as well as the Apple-approved audio formats.

Google's **Chromecast** is a screencasting dongle that plugs into the TV via HDMI. It streams audio and video from a tablet, smartphone, or computer running the Chrome browser through a home network to the TV. Android and iOS (via Safari) are both supported. Apps include Netflix, YouTube, and Google Play.

Computers as media and music servers

Computers can function as media servers with Windows Media Center (except in Windows 10) and various programs for Mac—the Mac Mini is a popular media server solution. For video recording, you need an in-the-clear signal without DRM and that can be hard to get from a cable or satellite operator. However, if you get good antenna reception, an **ATSC USB tuner** can feed a computer for recording purposes.

As music servers, computers are egregious audio abattoirs. Their constant multi-tasking mars the timing of the bitstream's zeroes and ones, causing audible degradation called **jitter**. Their digital-to-analog conversion is usually crude and loaded with noise. The best way to clean up computer audio is with a **USB DAC**. It can be **adaptive**, with the bitstream run by the computer clock; or **asynchronous**, which replaces the computer's clock with the DAC's clock. This can result in dramatically better sound. Most USB DACs are now asynchronous. High-end models are available from Meridian, Moon Audio, and others. Au-

151

dioQuest's DragonFly (sold in numerous versions for as little as $100) is a thumb-sized portable USB DAC. USB DAC capability is also starting to appear in top-end surround receivers from Cambridge Audio and Pioneer Elite. However, a USB jack on a receiver does not always accept direct computer input.

Introduction to audio file formats

- Lossy
- Lossless
- Uncompressed
- Bit depth/sampling rate
- CD quality
- Master Quality Recordings

Computer, network, streaming, and mobile audio devices use file formats that are classifiable in three basic groups. Some are also codecs, or encode/decode formats; others are just file formats.

Lossy audio codecs use heavy compression based on **perceptual coding** to eliminate some audio data that are masked by louder, more dominant audio data. They omit 80 percent or more of the original data. Some claim that listeners cannot tell the difference; trust me, you can, and the more data are discarded, the more objectionable the artifacts are.

Lossless audio codecs eliminate far less data, roughly 50 percent. They do not rely on perceptual coding and are higher in quality. While they do pack and unpack the data, in the end the original bitstream is reconstructed bit for bit, so the lossless file is as good as the uncompressed file from which it is derived. Lossless codecs are a good way to maximize storage capacity of a mobile device while maintaining quality.

Uncompressed audio formats do not encode, decode, or discard data. They are just containers full of bits. They are the least efficient in terms of storage, so for most network, streaming, and portable applications, lossless is a better alternative.

In addition to their file formats and encoding methods, audio files are defined by two characteristics. **Bit depth** refers to the number of zeroes and ones in a digital string. A 16-bit format (like CD) has a string of 16 zeroes and ones. A 24-bit format has a string of 24 zeroes and ones—which is not merely an additive difference, but an exponential one, so quality is much higher. **Sampling rate** is the number of strings transmitted per second. CDs have a sampling rate of 44,100 times per second, or 44.1 kHz. High-resolution files sold by retailers can have sampling rates of 48, 88.2, 96, 176.4, or 192 kHz. Note that 88.2 and 176.4 are multiples of 44.1, and 96 and 192 are multiples of 48.

A note about **CD quality**: This phrase, referring to 16-bit/44.1kHz audio, is often used as a synonym for high-resolution audio. My opinion is that the term is deceptive when used in that way because you can now download audio of much higher quality. High-res audio should have at least 24 bits, not 16, and a minimum sampling rate of 48 kHz. I refer to CD as a mid-res format. No one else does, but the distinction is useful.

To help consumers identify different kinds of high-resolution audio, the Digital Entertainment Group, Consumer Electronics Association, and Recording Academy have adopted the term **Master Quality Recordings**, which they define as "lossless audio that is capable of reproducing the full range of sound from recordings that have been mastered from better than CD quality music sources." They designate four kinds of MQ recordings: **MQ-P**, from a PCM master source, 20-bit/48kHz or higher, typically 24/96 or 24/192; **MQ-A**, from an analog master source; **MQ-C**, from a CD master source (16-bit/44.1kHz content); and **MQ-D**, from a DSD/DSF master source (typically 2.8 or 5.6 MHz content). Some of these designations may cause more confusion than they dispel. The specs for MQ-P set the bar too low: 20-bit audio is not high-res, though 24-bit is. And if MQ is supposed to be better than CD quality, what's the point of MQ-C? Whether high-res audio retailers will adopt this terminology remains to be seen.

Be warned that there's more to audio quality than just numbers. Content is only as strong as its weakest link. A great recording on CD will sound better than a terrible recording in 24/192 files. But a great recording will sound its best in the high-res form that the artist, producer, and engineer heard in the studio.

Lossy audio file formats

- MP3
- AAC
- WMA
- Ogg Vorbis

Though audiophiles spurn them, lossy formats are still popular for their greater efficiency. But they don't all sound alike. There are several codecs that provide better-sounding results than MP3 even when encoded at the same bitrates. They include Apple's **AAC** (Advanced Audio Coding, developed by Dolby Labs), the favored format of the **iTunes** online music store; Microsoft's **WMA** (Windows Media Audio), the favored format of the Windows Media Player; and **Ogg Vorbis**, used for Spotify streaming. But MP3 is the most widely used codec.

MP3 was originally named **MPEG-1, Layer 3** and was intended to compress audio on Video CD, a pre-DVD format. Napster and other file-sharing services subsequently gave it a new lease on life and made it famous. The inventor was the Fraunhofer Institute of Germany. Fraunhofer is a co-licensor of the format along with Thomson (former owner of the RCA TV brand). **MP3 VBR** is a slightly improved version of MP3. Conventional MP3 operates at a constant bit rate. MP3 VBR operates at a variable bit rate, depending on the music's demands. That allows it to provide slightly better quality in the same number of bits.

mp3PRO provides better sound at lower bitrates by providing a second stream of digital information. When mp3PRO files are played on an original MP3 player, the older player can access just one of those two streams. **MP3 Surround** adapts the popular stereo file sharing format to 5.1 channels of surround. It's backward compatible with existing MP3. These MP3 retrofits have yet to find widespread applications, and the music industry has shown no interest in them.

Lossless audio file formats

- FLAC
- ALAC
- WMA Lossless
- DSD/DSDIFF
- APE

The Ogg family also includes **FLAC**, or Free Lossless Audio Codec. Not to be confused with lossy Ogg Vorbis, FLAC has become the go-to codec for high-resolution music downloads, used by HDtracks.com and other retailers. FLAC can support up to 24-bit depth and (in practice) up to a 192 kHz sampling rate, making it greater than "CD quality."

The iOS and Mac platforms includes **ALAC** (Apple Lossless Audio Codec). It is associated with iTunes, QuickTime, and the AirPort Express wireless distribution device. Note that Apple's AirPlay wireless technology limited to 16 bits, or "CD quality." Microsoft's lossless codec is **WMA Lossless**, supported by the Windows Media Player.

DSD (Direct Stream Digital) is the high-quality lossless codec underlying the Super Audio CD. Its **DSDIFF**, or Direct Stream Digital Interchange File Format, is becoming an underground sensation among computer audiophiles, some of whom say it is even better than 24-bit FLAC. There are also double-rate, quad-rate, and octuple-rate versions.

The **APE** file format is encoded by the Monkey's Audio PC application and used in Cowan media players.

Uncompressed audio file formats

- PCM
- WAV
- AIFF

PCM (pulse code modulation) is the name given to uncompressed CD audio and to any form of audio with a bit depth and a sampling rate. The main equivalent in Windows is **WAV** (WAVEform audio). Apple's equivalent is **AIFF** (Audio Interchange File Format). Because they are less efficient than lossy and lossless codecs, they are rarely used in audio servers. Apple devices support WAV but usually the files are divorced from their metadata, making them difficult to access.

MQA: a new innovation in high-resolution audio

MQA, or **Master Quality Authenticated**, is a new method of audio encoding that tackles the challenge of making high-resolution audio more efficient while preserving full sound quality. It takes material recorded at a high sampling rate and uses various methods, nicknamed "audio origami," to fold down components less crucial to the ear into smaller amounts of data. Like lossy encoding, it allows smaller file sizes, and also allows high-res content to be transmitted at CD-quality data rates. But unlike lossy encoding, it causes no audible degradation.

Critics of high-res audio have pointed out that human ears don't need all the ultrasonic frequencies and extreme dynamic range of high-res formats. But the developers of MQA are more concerned with tiny temporal differences, or differences in timing, that they contend are detectible by human hearing as it has evolved over the millennia. Another advantage of MQA is that the DAC might be tweaked to compensate for the distortions that occur in mastering of original recordings. This raises the issue of provenance, which the music industry would have to tackle to fulfill this aspect of MQA's potential.

MQA is implemented in software and can be built into DACs, or digital-to-analog converters. It is available in Meridian products, having been codeveloped by Meridian's Bob Stuart, but can also be licensed to other manufacturers. Is MQA the next big thing in high-res audio? Only time will tell, but it is certainly provocative and promising.

Ripping

Of course you can always use your PC and an encoding program to rip your own music files from CDs or even analog sources. (Lots of my MP3s have LP surface noise!) The easiest way to encode MP3 or WAV is with **Windows Media Player** (version 10 and beyond), which rips at blinding speed, and does it without a plug-in. Of course, **iTunes** also does an excellent job with MP3 and WAV, adds the Apple-specific codecs, and sometimes has better metadata (for artist, song, etc.). For FLAC there are many options of which the best known is **WinAmp**.

Quality is an issue in **MP3 encoding** (and collecting). Like Dolby Digital, MP3 uses compression which does not provide a CD-quality (uncompressed or lossless) signal. The bare minimum for MP3s is 128 kbps (kilobits per second), though it doesn't sound very good. You'll get better results at 192 kbps, and 320 kbps approaches CD quality in most portable or casual listening applications. Some other encoders, like AAC and WMA, achieve higher quality at lower bit rates.

To rip from vinyl, you'll need a **phono preamp** with USB output to feed your PC. NAD, Pro-ject, and Bellari make inexpensive and good-sounding USB phono preamps. Another method would be to burn the vinyl to CD-R/-RW and rip the latter disc on a PC. To edit audio files on a PC, Audacity is an excellent choice, both versatile and free.

Metadata

Metadata are data about data. When your music player or server sorts material by artist, album, song, genre, etc., it's using metadata. **Gracenote**—once an open-source project, now a profit-making company—is the largest provider of metadata, and the favored provider for iTunes, but other metadata lookup services work with various products. The Windows Media Player works with **AMG** (the All Music Guide) and many other providers. **Musicbrainz** (musicbrainz.org) is a surviving open-source database.

When you rip from CDs with iTunes, WMP, or another encoding program, the program usually associates the disc with matching metadata. If you rip from cloned CDs or analog sources, you may need to edit the metadata to ensure that your player or server can find material by artist name, title, etc. You can do this manually, through iTunes, WMP, or even the Windows Explorer. Gracenote can identify a cloned CD using audio fingerprinting. For large metadata editing tasks, metadata editing programs can make the task easier.

Plagued by absence of album art in your music library? Your media player may leave unsightly blanks. iTunes often misses art for MP3 files not bought from Apple. Album art may be stored as a separate file in an album folder or, for most consistent results, embedded directly into each audio file. For MP3s, an ID3 tag editing program can help. Some tag editing or music management programs also embed images into FLACs. JPG is the most universally used image format for cover art. Images take up space, so keep file size down to 200 x 200 pixels (or 300 x 300 if you're more fussy about image quality than file size).

The Gracenote database and services

Digital audio servers that connect to the internet offer some significant convenience features. Many of them are provided by **Gracenote**. Now owned by Sony, this is the company that provides iTunes and an increasing array of music playback devices with metadata.

The Gracenote music recognition database (formerly **CDDB**) aggregates track metadata, 24 hours a day, from millions of users all over the world. It has downloadable data on millions of CD titles and tens of millions of song titles.

Gracenote's **MusicID** uses a fingerprinting technique to recognize not only whole CDs but individual songs in any format (MP3, WMA, AAC, etc.).

Gracenote **Mobile Music ID** identifies songs via cell phone. Hold your phone up to the radio and the database identifies the track. Then you can purchase the music as either a download or a ring tone.

Gracenote **Playlist** works in PC, mobile, and car products. Pick an artist or track as a "seed," and it generates and updates playlists dynamically using Gracenote's music similarity data and global popularity trends.

Gracenote **Playlist Plus** does the same for flash-based devices that may lack internet connectivity or memory.

Gracenote **Link** aggregates and delivers third-party content while music plays.

Gracenote **Music Enrichment** combines the sonic excellence of mp3PRO with the music recognition database, providing artist and title info, bios, photos, links, discographies, artist news, tour info, lyrics, and recommendations.

Finally, Gracenote **VideoID** organizes Blu-ray and DVD collections (in a changer or server) by title, release date, language, regional code, single/double-layer info, credits, running time, color/B&W, aspect ratio, and audio formats.

Connections for streamers, servers, and home networking

- digital audio input/output (optical or coaxial)
- stereo analog line input/output
- HDMI
- USB
- ethernet network connection
- wireless network connections
- phone line network connection
- powerline network connection
- iOS docking/USB connections
- analog headphone connection

Streamers, servers, and mobile devices connect to your system like any other digital component, using digital optical or coaxial connections, stereo analog line connections, and (in media servers) HDMI and various other video connections.

A computer can connect to an audio system using its analog or HDMI outputs. However, a USB connection used in conjunction with a high-quality USB DAC can provide better results.

Network-connected devices may use ethernet or wi-fi. Ethernet is easy to arrange in new home construction; retrofitting an existing home is harder but not impossible. The most common kind of home network cable uses **internet protocols** and a **router** (to split and distribute the signal). The wiring is usually **Category 5** with **RJ-45** connectors. Cat5 cable carries the equivalent of four separate **twisted pair** (phone) lines though it can distribute various kinds of signal. A variation called **Cat5e** has a center spine that separates the four twisted pairs, reducing crosstalk, and therefore allowing greater bandwidth which translates into higher-speed data transfer.

However, especially in older homes where new wiring is an intimidating task, there are alternatives. Many products support **wireless** signal-transfer schemes such as the **802.11b** standard (better known as **Wi-Fi**). The *b* version operates at a pokey 11 megabits per second—plenty for MP3 but not for video. **802.11a** is five times faster, at 54 Mbps, but is not compatible with the *b* version. However, **802.11g** operates at the same speed as *a*, and is backward-compatible. **802.11n** transmits up to 540 Mbps at distances up to 160 feet. **801.11ac** transmits up to 1.3 gigabits per second using the 2.4 GHz and 5 GHz bands to reliably stream HD video. It is available in routers and will eventually reach smart TVs and other network video devices. Until that happens, you can use an **AC bridge** to get a/v signals from router to TV.

What's beyond that? Possibly **802.16**, a.k.a. **WiMAX**, a broadband wireless access standard which extends range from 300 feet to several miles. Clearwire is an example. It may someday replace wired broadband connections (or at least provide some stiff competition). A wireless version of **IEEE 1394**, also known as FireWire, was finalized in 2004 and offers an attractive blend of high bandwidth and convenience.

Products with HomeGrid's **Home PNA** standard use **RJ-11** phone jacks and existing phone wiring to spread networked audio throughout a home. The phone line can be used for calls at the same time as long as the line isn't being used for a dial-up net connection. Home PNA 2.0 has a maximum speed of 10 megabits per second (more than enough for MP3 data rates). The **Home PNA 3.1** standard increases the data rate over phone wiring to a speedy 320 Mbps over two simultaneous channels, enough for compressed video. For more information on the Home Phoneline Networking Alliance visit homepna.org.

Another non-invasive method of networking called **HomePlug** uses existing home power lines for audio, video, VOIP phone, and home networking control. The latest version, **HomePlug AV2**, has improved throughput speed, transmission range, and operating spectrum. You can use it to network a broadband connection, multiple 1080p video streams, 3D or UHD video, security cameras, and internet gaming. For more on the HomePlug Powerline Alliance see homeplug.org.

Finally, iOS devices (iPhone, iPad, iPod) have become audio/video servers in themselves. While wireless Apple AirPlay and Bluetooth are becoming the preferred connections, you might also use the older 30-pin

iPod docking connector built into older iOS gear (except the iPod Shuffle) or the newer **Lightning connector** in newer gear. Some of these devices are docks that connect to surround receivers, compact systems, or even multi-zone audio systems. Some are standalone systems in themselves, with their own drivers and amps, that use iOS as the key (though not necessarily the only) signal source. A select few docking devices offer a high-quality digital connection. Others use an analog line connection, which still sounds better than the player's headphone output. Note that the iPhone's new eight-pin Lightning connector can function with 30-pin accessories only with an adapter. In addition, an iOS device can connected to some surround receivers using a direct USB input with the Apple-supplied cable.

However, any tablet, smartphone, or music player can connect to a system using its **headphone output**, which is basically a volume-controlled analog stereo output. You'll need to get an adapter cable with a 1/8-inch mini-plug at one end and two RCA-type plugs at the other. Manufacturers like BetterCables.com and Wireworld make very good ones; if you're desperate, there's always RadioShack.

Satellite receivers

Monster dishes that dominated a lawn have almost completely given way to small dish systems that sprout from the roofs and windowsills of both suburban and urban homes. Both of the major small dish systems link an 18-inch dish to a rack-sized receiver box. (Residents of Alaska may need a larger dish.)

DirecTV Satellite System (now owned by AT&T) and **EchoStar's Dish Network** are both small-dish systems that bring anywhere from 100 to more than 200 channels of programming including pay-per-view movies and events. The hardware is generally cheap while the programming is competitive with cable TV rates in most areas. Both offer HD programming. And both have just begun offering Ultra HD.

Picture and sound quality

Like Blu-ray and DVD, satellite video uses digital compression to efficiently provide a high-quality picture. The quality of that video varies according to atmospheric and other conditions, such as channel-crowding on the satellites.

The word *digital* does not automatically signify superiority. Digital satellite video does not always attain the quality of Blu-ray or over-the-air HDTV. Picture quality can look marvelous on a good night, terrible on a bad one.

Unfortunately, bad nights have become more frequent. The constant pressure to add new channels to the finite number of satellites in orbit has reduced overall picture quality. New choices are great, but unless they're accompanied by enough new bandwidth, something has to give. A picture filled with blocky-looking digital distortion has become the bane of the keen-eyed satellite viewer. HDTV over-the-air broadcasts generally look better than satellite- or cable-delivered HDTV.

Both DirecTV and the Dish Network offer **Dolby Digital** surround sound on some premium movie channels as well as selected pay-per-view movies and music events; DirecTV also supports **Dolby Digital Plus**. You'll need a Dolby Digital-compatible receiver to get it.

Recording options

If you'd like to use your satellite service with a **DVR** (digital video

This DirecTV 5LNB satellite dish has the five LNBs (or internal antennas) necessary to get HDTV. It is distinguished by the word "Slimline" printed on the dish.

161

recorder) or **PVR** (personal video recorder), DirecTV and the Dish Network offer satellite receiver packages with built-in hard drives, so you can pause the program while raiding the fridge. High-def DVR recording is also available from DirecTV/TiVo and the Dish Network.

The Dish Network's **Hopper DVR** has attracted the animosity of networks thanks to its ad-skipping feature, **AutoHop**. Hopper also uses IP video to send programming to your computer, tablet, or smartphone anywhere there's an internet connection. And its PrimeTime Anywhere feature makes available three hours of primetime programming up to eight days after air date. To extend the Hopper's output to other rooms, Dish offers a local receiver called **Joey**, connected via coaxial cable. A new version called **4K Joey** handles Ultra HD.

Ultra HDTV via satellite

Both of the major satellite operators have recently announced they would begin delivering Ultra HD. DirecTV offers the **Genie DVR** (see directv.com/technology/4k) and Dish Network offers the **4K Joey receiver** (see dish.com/4k-joey). Details were sketchy at presstime but Dish Network said its version supports new HEVC H.265 video compression as well as older MPEG-4 and MPEG-2, a 60 Hz frame rate, 10-bit color depth, and network features via the Hopper or Hopper with Sling DVRs. The picture-quality features are a mixed bag: H.265 can carry an assortment of next-generation enhancements, while MPEG-4 and MPEG-2 are existing HD technology. How good satellite UHD will look remains to be seen. But we all have to start somewhere and the satellite operators need to do a good job with UHD to stay competitive with the likes of Comcast, the first cable operator to announce UHD.

HDTV via satellite

Getting high-definition television by satellite requires a different-shaped dish than the conventional 18-incher to receive HDTV. DirecTV's HD broadcasts originate from several satellites, so they require a satellite dish with multiple **LNBs** (internal antennas). The newest 5LNB dish has the word Slimline printed on it. You'll also need an HD-capable receiver or DVR plus a **B-Band Converter** to access new HD channels. DirecTV now supports 1080p video with the right hardware.

DirecTV offers HD versions of channels plus DVR service (well worth the few extra bucks). The selection of packages for sports fans is huge. But when you hear DirecTV make claims for HD offerings, be advised that includes satellite radio channels, regional networks not available in all areas, duplicate feeds of certain networks, and other fluff. What most people would regard as the core offerings are more like a couple of dozen channels. Local HD channels are delivered separately by antenna in some areas due to legal restrictions. However, following negotiations with local stations, DirecTV says it now reaches 94 percent of U.S. households with local channels.

DirecTV's HD image has fewer pixels than over-the-air HDTV. A 1080i-format HDTV picture normally would have 1920 by 1080 pixels, for a total of 2,073,600 pixels. DirecTV provides fewer horizontal pixels in a 1280 by 1080 picture, totaling 1,382,400 pixels—nicknamed HD Lite. The picture is not cropped, just lower in resolution, and the difference may not be noticeable on all HDTVs. However, by switching from MPEG-2 to MPEG-4 video compression, DirecTV now supports 1080p on some channels. The Dish Network's HD feed is not truncated.

The Dish Network also offers HD. Local channels no longer cost extra; they are bundled, along with DVRs, in certain packages. The Dish Network's two HD set-top boxes come with a built-in antenna to bring in local off-air DTV stations. These channels also appear in the program guide and can be recorded by the Dish HD DVR Receiver.

Both major satellite operators market internet access of various kinds through partners. See their websites for more details.

DirecTV and EchoStar (operator of the Dish Network) change their programming lineups and pricing frequently. For the most up-to-date information see directv.com and dishnetwork.com.

Other features

Dual LNB is the name given to satellite receivers that can serve up more than one channel at once. The acronym stands for **low noise block downconverter**. Models with this feature may cost slightly more. You don't need dual LNB to feed more than one TV if all screens display the same signal. Likewise, if you receive a pay-per-view event, and want to feed it to more than one TV, there's no extra charge.

Parental lock can prevent children from viewing adult material.

Connections

- RF satellite input
- RF antenna input
- RF output
- HDMI output
- component video output
- S-video output
- composite video output
- stereo audio outputs
- phone jack
- Dolby Digital output (optical or coaxial)
- data port
- RF remote input

The satellite feed enters your system through an **RF satellite input**. To provide local channels, some of which may be missing from your local satellite feed, there's also an **RF antenna input.** At one time satellite operators were forbidden to provide any local channels, due to federal regulations, but in 1999 Congress began easing these restrictions, following viewer outcry.

The **RF output** will feed older TVs via channels 3 or 4, but the best picture quality will come from the **HDMI** or **component video** outputs. Ultra HD requires HDMI 2.0/2.0a. **S-video output** is the next best choice, while **composite video output** is the third.

Stereo audio outputs may carry Dolby Surround, which can be decoded in Dolby Pro Logic, but a better choice is the **Dolby Digital output**, usually the optical type. Once a relative rarity—early generations of DirecTV (then called DSS) satellite gear didn't have it—Dolby Digital has become standard equipment.

The **phone jack** provided on all satellite gear allows the system to dial into the satellite operator for billing purposes. The **RF remote input** accepts a small antenna that makes it possible to use a remote that penetrates walls with radio frequencies (as opposed to the weaker infrared signals that come from most remotes).

Antennas

Digital broadcast television ... public radio and talk radio and Top 40 radio ... this is the glorious low-rent entertainment that gushes into our home theater systems via antenna. Over-the-air reception is the first, sometimes the best, and sometimes the only way to get digital TV. It's also the only form of DTV that's free.

While most Americans get television from satellite or cable providers, the antenna is the medium of last resort for outlying areas that aren't wired for cable. Satellite broadcasters are now required to deliver the major networks in the top 20 metro areas, but still do not deliver all local channels in all areas. The FCC adopted "must carry" rules in 2007, ruling that cable systems must deliver all local channels to all sets, digital or analog. But not everyone can get cable. Thus an antenna may be the only way to get DTV with at least some local channels.

The DTV transition brought analog broadcasting to an end on June 12, 2009. If you depend on over-the-air reception, digital is now the only kind that works. To adapt an analog TV to digital broadcasting, you'll need a set-top box. To feed a DTV with digital signals, you may need a different kind of antenna than before. That's because your new digital signals may be traveling the airwaves over a different set of frequencies. To find listings of local channels, go to dtv.gov or tvfool.com.

TV channels 2-13 are **VHF** (very high frequency) while channels 14-83 are **UHF** (ultra high frequency). The **lowband** channels, from 2 to 6, are more susceptible to noise than the **highband** channels, 7 to 13. As a result, stations that decided to stay on channels 2 and 6 in the DTV era are having trouble reaching viewers. Channel 6 seems to be especially problematic. Between the lowband and highband channels (6 and 7) is the entire **FM radio** band.

Digital channels may be either VHF or UHF. The DTV transition has caused a game of musical chairs. At the start of the transition, while analog channels were still operating, the Federal Communications Commission assigned unused UHF slots for digital channels. Stations operated at both frequencies for several years, providing both analog and digital service. However, as the analog VHF channels were vacated in 2009, some of the digital channels moved into VHF—enabling networks to reoccupy what they and viewers regard as their traditional slots. The lowband VHF slots, traditionally noisy, are less desirable and some stations are abandoning them for higher frequencies.

Size and range

You're in luck. Smart, well-organized people are trying real hard to make sure you buy the right antenna, with the size and shape that work best in your area. They've provided a color-coded map that'll tell you just what you need. Look for **AntennaWeb** at antennaweb.org, compliments of the Consumer Technology Association.

Bigger antennas are for those who live farthest from the transmitter, or who have other reception-strangling problems stemming from local conditions or long cable runs. Those who live in or near major metro areas can get by with a smaller antenna. Bigger is not always better: too much signal may overload the TV tuner. Rather than get hung up on size, focus instead on the antenna's **rated range**, and let the AntennaWeb map be your guide.

Some of the biggest antenna problems are easy to avoid: Splitters should be used sparingly. And attic placement always cuts signal strength by at least half, even in wood-frame construction. A larger antenna might or might not help. With materials other than wood, the problem worsens. Aluminum siding stops signals cold. Other signal inhibitors within your home's walls may include the chicken wire inside stucco, the copper shielding inside terracotta, and brick. If roof mounting isn't practical, consider some other outdoor location.

Local conditions can affect how well an antenna works as much as range. Tall buildings, power lines, lighting rigs, and hilly terrain all may contribute to reception problems. One of these problems is **multipath distortion**. The term literally describes signals taking multiple paths to the antenna and distorting the picture. When a signal hits a tall building or other obstruction, it splits into pieces. The antenna receives both the main signal and fragments of it that arrive a split-second later. In analog TV, this caused ghosting. Digital TV doesn't have ghosting problems, but multipath distortion still may prevent the set from getting a usable signal. A signal fragment more than 50 percent as strong as the original signal will prevent DTV signals from getting through at all. A more directional antenna will help minimize multipath distortion.

TV antenna types

Outdoor antennas pull in stronger signals than indoor ones, except when mounted in the attic—then they *become* indoor antennas. For outdoor use, pick an antenna based on your reception needs, not cosmetics.

For **indoor** use, you have alternatives, starting with the **dipole rod** and **UHF loop** supplied with smaller TVs, a sculpted designer antenna, or traditional rabbit ears. For FM, a designer antenna may look better than a simple (and unsightly) wire antenna, but may not sound better.

Yagi (or **Yagi-Uda**) is the name given the most common type of outdoor TV antenna. Its interconnected rods, or **elements**, are of different lengths to capture the different frequencies that correspond to each channel.

Directional (or **unidirectional**) antennas receive signals from transmitters lying mostly in one direction. This may be a better choice if your local transmitters are all part of the same metro area. Again: consult AntennaWeb.

Multi-directional (or **omnidirectional**) antennas receive signals from all or most directions equally well. This is a better choice if you live between various transmitters.

Motorized antennas include a mechanism that enables the antenna

The best way to get HDTV is over the air. A Yagi-type antenna like the Channel Master CM 2016 can receive digital channels in both the VHF and UHF bands.

to rotate, facing in different directions to receive channels from transmitters in various locations.

Active antennas boost the signal with built-in amps. Most antennas are **passive**, without amplification, though it can always be added if needed. Just be advised that boosting the signal will not correct defects in the signal (and may overload the tuner).

Satellite dishes sometimes come with built-in antennas to supplement satellite channels with over-the-air channels. There are also antennas designed to mount alongside satellite dishes.

DTV antennas

DTV operates in both the UHF and VHF bands, so antennas that work with those signal types work for DTV. People in most areas will need both. Some of these antennas are being labeled DTV or HDTV antennas.

UHF signals have a shorter wavelength than VHF signals and therefore are captured by shorter rods. It is not unusual for a Yagi antenna to include both the shorter rods that capture UHF signals and the longer rods that capture VHF signals.

Another UHF antenna type is the **bowtie,** whose distinctive triangular rods can be arranged in multiples to pull in signals more effectively. The simplest UHF antenna is the loop supplied with small sets or built into rabbit-ear antennas.

The good news about DTV antennas is that DTV tuners—whether they're separate boxes or built into sets—have seen marked improvement in recent models. That should make any antenna work better.

Radio antennas

FM radio lives in the middle of the VHF-TV band, between channels 6 and 7, so any antenna that works for VHF television will also work for FM radio. To radio lovers, this is an opportunity, but to videophiles it's a problem, so some antennas (and cable systems) will strip out the radio signals with **FM traps** to prevent interference with TV channels 6 and 7.

An FM antenna takes many forms. It may built into the radio, or its power cord, or may be a piece of wire supplied with an audio/video receiver, AM/FM tuner, or table radio. Some wire antennas are Y-shaped to pick up stations in different directions. Some of the most attractive designer antennas are FM antennas.

DTV broadcasting vs. mobile broadband

Free over-the-air TV may be an endangered species. Antenna-dependent households are far outnumbered by those subscribing to pay-TV, at just 17 percent of U.S. households in 2016 according to market researcher GfK. However, the number is up two points from 2015, with cord-cutters switching from cable or satellite TV to broadcast TV. And other users are greedily eyeing the DTV spectrum. Despite having allocated spectrum to DTV in a decades-long process, the Federal Communications Commission is seeking to reallocate much of it to mobile broadband to suit the nation's latest priorities. While fostering mobile broadband is a good thing, shrinking availability of DTV signals to viewers may be a bad thing, both for low-income viewers dependent on antennas, and for videophiles who prefer a high-quality HD ATSC signal to low-quality video streaming. While the FCC says its spectrum reallocation auction won't reduce broadcast TV service, it remains to be seen how many stations will remain available in any given area and at what signal strength. It's all one big science experiment and TV viewers are the guinea pigs.

The next DTV transition

The U.S. completed the transition from analog TV broadcasting (NTSC) to digital (ATSC 1.0) in 2009. Another transition, to ATSC 3.0, is planned to accommodate Ultra HD and other new technologies. How it will interact with the spectrum auction remains to be seen. For details, see "Television/UHD and 3DTV/ATSC 3.0: UHD over the air."

Useful Accessories

S ome accessories are afterthoughts. Others are bubble-packaged plastic marketing scams. But the right accessories, regardless of profit margin, are those that help your system achieve better performance or greater convenience. They enhance the value of what you already own and make it more pleasurable to use.

Remote controls

Nearly every time you add a component to your system, you add a remote control. After a while they're collecting in a basket, or filling up drawers, or just scattered all over the place. At that point you wonder, wouldn't it be great if one remote could do it all? A good remote may not handle every single function, but will give you the basics—source switching, volume, mute, power. More sophisticated remotes really can do it all. Perhaps, though, a better goal would be to find a remote that'll just make life easier.

Universal remote control

A true **universal remote control** handles more than one brand of product. It may be packaged with an audio/video receiver, or some other product, or may be bought separately.

At its simplest, a "universal" remote may simply control other products of the same brand. Your TV may come with a remote that controls the same brand's disc players. That will work as long as you prefer to stay within that brand.

Thanks to HDMI, control automation has taken a further step forward with **CEC (Consumer Electronics Control)**. It took effect with HDMI version 1.2a. Compatible DTVs, receivers, disc players, etc. communicate and coordinate commands as described above. Unfortunately manufacturers are muddying the waters by using their own proprietary names for this non-proprietary technology—such as Panasonic's Viera Link, Sony's Bravia Theatre Sync, Sharp's Aquos Link, Toshiba's Regza Link, etc. It's all CEC-based though interoperability among brands is not guaranteed. There are also some older pre-CEC control systems that enable components to jointly respond to commands.

If, like most home theater buffs, you end up with a multi-brand system, the most useful universal remote is the **programmable** kind.

Preprogrammed vs. learning remotes

There are two kinds of programmable remotes: preprogrammed and learning. Some remotes do both. Each kind acquires command codes in a different way. They may be bought as separate products or, if you're lucky, supplied with certain products such as receivers or TVs.

A **preprogrammed** remote has a built-in "library" of codes for different makes and models. Depending on the model, the database of codes may be updated to control products you add to your system in the future. A preprogrammed remote typically covers the most popular brands and models. If you think your TV brand is too obscure, you'd be surprised—there may be a preprogrammed remote that covers that brand. However, high-end audio products such as stereo preamps are

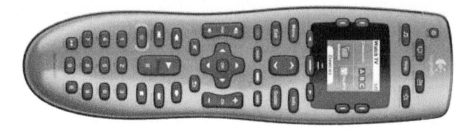

The Logitech Harmony 650 ($70) makes it easy to set up one remote control to rule them all (your components, that is). Instead of an elaborate button-pushing routine, this remote lets you enter information about your components into a browser-based form. Then download the commands into the remote via USB and you're in business.

171

not as well covered.

A **learning remote** acquires codes directly from the original re-motes that came with your TV, disc player, etc. When placed end-to-end with an original remote, it "learns" commands. The original remote emits the codes and the learning remote accepts them. A learning re-mote is inherently updatable—that's what makes it a learning remote. If you have a lot of legacy equipment, or plan to buy new equipment in the future, a learning remote will handle all those products.

Some remotes have both preprogrammed and learning capabilities. The preprogrammed side will speed setup, while the learning side will provide needed flexibility for special programming problems. A few learning remotes include some preprogrammed codes, but only for one brand (example: Marantz).

But most universal remotes are either preprogrammed or learning. Neither type is automatically superior. A preprogrammed remote re-quires you to enter the numeric codes that identify each of your compo-nents to the remote. You may be able to punch them in; or you may have to cycle through every code in the remote's internal database be-fore finding one that activates the component.

A learning remote takes longer to set up. You're not telling the re-mote to recognize a make and model—rather, you're individually trans-ferring each command that you intend to use. You'll quickly discover that transferring them all is rarely worth the effort and that front-panel controls have their uses. You may have to transfer some codes more than once to get them right.

Number of controlled components

A universal remote may control a small system or a large one. Get one that will handle all remote-controlled components in your system. For a small bedroom system with only, say, a TV and cable box, a low-end remote that handles those devices will be adequate. But it will not cover a system that also includes an audio/video receiver and a rack stuffed with other components. Moderately priced universal remotes handle 8 to 10 devices. The pricier ones do well over a dozen.

Look and feel

Whatever kind of remote you buy, it should **feel** good in your hand. Rounder shapes typically have a better feel. However, the traditional flat shape lets you rest the remote on the arm of your chair, so you can punch the keys without having to hold the remote.

Button size involves a tricky tradeoff between simplicity and flexibility. When buttons are larger and fewer, the remote is easier to use—but can't do as much. When buttons become small and numerous, you can do more, but with greater difficulty. There are also some remotes that attempt to do a lot with a few buttons, relying on LCD screens. These can be wonderful or maddening.

Controls should be easy to tell apart by **size, shape, color, background color, layout,** or **labeling.** The most oft-used controls should be biggest. Volume controls should be large and easy to find. Certain traditional shapes work well for transport controls (play, pause, etc.). And there's no reason why all buttons have to be grey or black. Manufacturers who drown our fingertips in a sea of identical black-on-black buttons really need to invest a few more pennies in colored plastic.

Control layout also affects ease of use. Your fingers or thumb should easily find the volume up/down and mute buttons. Commonly used functions should not sit too close to controls that stop the show. Certain control settings belong under a lid, where they won't respond to an accidental touch.

Step-up remote features

Glow in the dark or **luminescent** keys make controls easier to see in a darkened home theater. They may be **backlit,** with a light source beneath the keys that glows briefly. The light source may be switched on manually (with a backlighting key, preferably side-mounted) or automatically (whenever you touch a button or the remote senses vibration), giving you just enough time to make up your mind and hit a key. **Phosphorescent** keys, like the hands of a watch, absorb room light and keep glowing after the lights go out.

Macro keys—a top-of-the-line feature that has trickled down—store whole strings of commands. You might set up one macro to turn on your receiver, select the BD input, power up the BD player, and extend its disc tray. Another macro might power down the whole system.

RF remotes are as handy as they are rare. They use radio-frequency signals in lieu of the infrared signals used by most remotes. RF signals can penetrate walls. Even when used within the same room, an RF remote is likely to have greater range than an infrared remote.

Touchscreen, PC-compatible, and app remotes

Some higher-end remotes come with an **LCD touchscreen**. Unlike the liquid crystal display on your calculator or watch, a touchscreen responds to the touch of your fingertips. These so-called **soft keys** provide lots of extra flexibility. You can add or remove buttons, move them around, or rename them. You can even relabel them with icons. The remote becomes a willing servant that does whatever you want (if you're willing to work at it long enough).

The most versatile remotes are **PC compatible**. Some accept commands from the internet or a PC. The slickest ones, from Logitech, provide a web interface that focuses not on individual commands but on how you use your system. If you watch a Blu-ray disc by turning on your TV, receiver, and BD player, the interface will determine the necessary commands and the remote will download the commands. Then you can start a viewing session with one button. By substituting computer literacy for consumer electronics literacy, which is rarer, web-enabled remotes have made it easier to get one remote to do it all.

The newest idea is to use multi-purpose handheld devices as wi-fi enabled remote controls. Most surround-receiver manufacturers provide free **apps** for the iPhone, iPad, iPod touch, or Android phones. Sometimes they even work better than conventional remotes. The iControlAV2 app for Pioneer receivers can adjust side-to-side or front-to-back audio balance by tilting the iDevice.

The best high-end remotes, once you've massaged the fine points of programming, make your system easier and more fun to use.

Upgrading and maintenance

You can always upgrade a learning remote. It's just a matter of placing it end-to-end with whatever remote comes with a new component. Transfer the codes and the original remote can go in a drawer with the others. (Keep them in case you need to program a new learning remote.)

Upgrading a low to moderately priced preprogrammed remote may be possible in some cases. Some will accept new codes via CD-ROM or download. Others have to return to the factory to receive new **firmware** (software built into chips). Don't expect all cheap universal remotes to be upgradable.

If you need to change the batteries, don't worry about losing your programming. Most remotes come with a **battery backup** lasting from

90 seconds to 15 minutes. The remote will retain its codes long enough for you to replace the old batteries with new ones.

Power-line accessories

Products that mediate between house current and audio/video components take three forms: power strips, surge suppressors, and power-line conditioners. A power-line accessory may combine some or all of these functions.

Power strips

These simplest of power-line accessories do nothing more than convert one outlet to several, possibly with a fuse in between. Do not use a power strip in a home theater system. It will do nothing to protect your components, may inhibit their performance in some cases, and is often so poorly made as to constitute a fire hazard. (A tenant in my apartment house once caused a fire by using a power strip. I watched smoke and flames pour out of his windows. He left the building on a stretcher and was later evicted.) It's better to plug a source component or two into the back of your receiver than to use a power strip.

Surge suppressors

It's better still to use a *high-quality* surge suppressor. (Please note that the term "surge protector" is technically incorrect—a surge is not something to protect, but something to protect against.)

Good ones will protect your equipment from potentially damaging **surges**, **spikes**, or **transients** in the 120-volt power supply. That includes not only lightning strikes but more mundane power-line disturbances like an air conditioner's compressor switching itself on and off. Anything with a motor—a printer, copying machine, hair dryer, vacuum cleaner—will temporarily destabilize the power supply, possibly causing damage to delicate circuits in home theater components.

Most surge suppressors use **metal oxide varistors**. **MOVs** are semiconductor parts whose resistance drops as voltage rises. That enables them to soak up excess electricity and send it to ground. More than

175

a hundred small surges per month can occur in an average household, so your power-line accessory's MOVs have their work cut out for them.

Some new products, including those from Tributaries, use **silicon avalanche diodes**. Unlike MOVs, which fail following repeated jolts, SADs send overvoltages to ground and trip a fuse. So instead of junking the whole product, you change a fuse.

When shopping for a surge suppressor, first look for **UL 1449** certification. This indicates that the product has been tested by Underwriters Laboratories for surge suppression (using 6000 volts of electricity!) and will not burst into flames if overheated. Because UL-certified products vary in performance—UL checks performance ranges, not specific benchmarks—you have more homework to do. There is also a **UL Standard 6500** certification which ensures that the product complies with all safety requirements of the **National Electrical Code**.

Next, check out the **joule** rating, a measure of how much electrical energy the product can absorb before it fails. You'll see a lot of three-figure numbers; for a home theater system, look instead for a four-figure number. And take it with a grain of salt. Some manufacturers make exaggerated claims concerning joule ratings. These ratings still don't tell the whole story.

Inflated joule ratings have led to an alternative rating known as **voltage let-through**, a measure of how much voltage the surge suppressor will conduct before it starts clamping. The less, the better. Even this specification doesn't tell the whole story because it fails to describe the shape of the surge's waveform. But the best choice for your system will have both a high joule rating and a low voltage let-through relative to other options.

Speed is also a factor. Look for some mention of **response time**. This is a measure of how quickly the product can react to protect your components. A response time of one nanosecond or less is good. A nanosecond is one-billionth of a second.

Power-line conditioners

Good power-line accessories—especially those calling themselves **power-line conditioners**—go further than just protecting your components. They filter out interference, isolate components from one another, and have status indicators for polarity and ground, enabling your components to perform at their best.

RFI or **radio-frequency interference** is an impediment to good

176

performance. Basically, every piece of wire in your system—including interconnects, speaker cables, and power cords—can act as an antenna that picks up radio, TV, and other signals. Good power-line accessories filter RFI out of the power line before it reaches your components.

EMI is **electromagnetic interference**, caused when components radiate noise, either through the power line or through the air. The best power-line conditioners have **isolation transformers** to prevent components from polluting other components through the conditioner.

Sequential turn-on is a useful feature when the power-line conditioner is used in a large system. By slightly staggering the powering up of each component with a slight delay, it provides two benefits. First, it prevents a sudden massive power draw, with several components powering up at once, from tripping circuitbreakers. Second, it prevents switch-on noise from being heard through (or damaging) the system. Some components have circuits that make noises when they wake up. By placing your power amp after such components in the turn-on sequence, you can avoid that noise, sparing both speakers and ears.

Because electricians are human and can make mistakes, **indicators** are useful features. They may tell you whether the electrical outlet is correctly **polarized** (with the positive and negative elements in correct position) and **grounded** (to conduct electrical shocks safely into the earth).

Polarity is the reason why one blade of a two-prong plug is wider than the other. Inverted polarity isn't necessarily a safety issue but can cause performance problems. Absence of ground—the third rounded prong in a three-prong plug—can be a safety issue.

Improper grounding can also cause a performance problem known as a **ground loop** when power and interconnect cords inadvertently set

The Power Wedges from Audio Power Industries include isolation transformers to prevent components from polluting one another with interference. Shown is Model 1118 ($1449).

177

up a closed circuit between two unevenly grounded components. You may see the problem as a large noise bar rolling through the picture. Or you may hear it as a distracting hum.

High-end power-line conditioners (from companies like Audio Power Industries, Tributaries, and Monster) can be quite costly. But that's because the best ones use hospital-grade circuitry to ensure that the performance of your costly components won't suffer. (Do athletes consume nothing but white bread and polluted water? Do world leaders dine on baloney sandwiches? 'Nuff said.)

Voltage stabilizers

A new and potentially helpful kind of power-line conditioner is the type variously known as a **voltage stabilizer**, **AC regenerator**, or **AC regulator**. These devices store power and ensure that connected devices receive a steady 120 volts. (That's for the U.S. The electrical systems of some nations, for example the U.K., differ.)

In the best of all possible worlds, that's exactly what your AC outlets would deliver. However, homes are situated varying distances from local power stations and those farthest away may receive a range of chronically inadequate voltages. When the power company is experiencing peak demand—routine in summer or winter—or damage to its plant, it may deliberately reduce voltage across the board, resulting in **brownouts**. Closer to home, someone activating a power-sucking appliance in your home, or even next door, can cause a **power sag**.

These power-line fluctuations can cause performance problems. When a video display gets less juice than it needs, its power supply struggles to keep up, and the image may show flicker or video artifacts. Likewise a surround sound system experiencing a power drain may be subject to a sudden shrinkage or collapse of its soundfield.

To determine whether a voltage stabilizer would be helpful in your system, perform periodic spot-checks of your home's power supply using a **volt meter**. You can find an inexpensive one at RadioShack.

Uninterruptible power supplies

To protect against **sags** in the power supply, look into products that have an **uninterruptible power supply** (**UPS**) with **automatic voltage regulation** (**AVR**). They can prevent power from falling below or rising above a fixed level. These features are more critical for high-end computer users than for home theater systems. However, as audio/video products acquire more and more delicate computerized parts, UPS and AVR are starting to look like good ideas.

Correct use of power-line accessories

Don't plug a receiver or power amplifier into a surge suppressor or power-line conditioner unless it has a receptacle designed specifically for amps, which draw higher current than other components. Otherwise the dynamics of your system may suffer. Well-designed products will take this into account and provide one or more appropriately labeled outlets for receivers or amps. If you're a finicky audiophile, you might want to arrange loans from your local dealer, and listen to the sound of your amp before and after inserting prospective power-line conditioner purchases, to find what works best with your equipment.

Many surge suppressors come with insurance that will reimburse you if the surge suppressor fails and allows *properly connected* components to be damaged—note the proviso in italics. Every power cord, every video connection (antenna, satellite, cable), every phone connection (like the one on a satellite receiver), and every network connection (receivers, Blu-ray players, streamers, servers) must go through the surge suppressor. Appropriate products will have receptacles for all these different kinds of connection. (Panamax products have a modular construction to allow subsequent add-ons.)

While a surge suppressor can handle the consequences of a nearby lightning strike, even a good one cannot defend against a direct hit. For maximum protection, make sure there's a **lightning arrestor** on your antenna, satellite, and cable feeds. Don't assume the installer will automatically provide it—some may neglect this important precaution.

Failing to make all the right connections is like walking out of your house and leaving the front door unlocked. It takes only one slip-up for an electrical intruder to get in and wreck the place.

Cables

A home theater system requires several kinds of cable including speaker cable, video interconnects, and audio interconnects (digital or analog).

Do you need high-end cable? Our motto here at *Practical Home Theater* is *be practical.* Experiment and use what works best. You may luck out and find a dealer who will loan you cables for evaluation.

It is true that speaker cables and analog interconnects can sound different from one another. (Electrical characteristics such as resistance, impedance, and capacitance account for most of the differences. Unfortunately, companies marketing high-end audio cables almost never specify these things.) But don't rush into any rash, expensive purchases when you're setting up new speakers or amps and just getting to know them.

A lot of jargon surrounds the marketing of premium cables. You may see a lot of technical terms such as **dialectric** (insulation between conductors) and **skin effect** (current-flow pattern that skews frequency response). Manufacturers may boast about materials (oxygen-free copper is common, silver less so) and construction (the way the strands are wound or braided). While these things hold great fascination for advanced techies, most people are better off concentrating on the basics such as thickness, insulation, shielding, termination, and fireproofing.

If you're running cable behind walls, use the **fireproof** CL2 and CL3 designations, because cables can act as fuses, spreading fire. Consult a CEDIA technician and/or get a local electrician to pull cable.

Regardless of whatever kind of cable you're working with at the moment, try your best to minimize the length of all cable runs. Long cable runs degrade video and audio signals alike.

Speaker cable

The **thickness** of speaker cable is measured in **AWG**, which stands for **American wire gauge**. Speaker cable is most practical when it is generic and thick (at least 16 AWG, and as much as 12 AWG). As the gauge number drops, cable thickens. Thicker cables are more appropriate for the longest cable runs such as those to the surround speakers because they minimize signal loss.

One benefit of premium cable is **insulation**, to provide some measure of protection from abrasions, and **shielding**, to keep the signal from being polluted by **EMI**, or **electromagnetic interference**, from nearby components. If your system can receive a watchable picture from an indoor TV antenna, or a clean FM stereo signal, it may also be subject to **RFI**, or **radio frequency interference**, as cables can literally act as antennas, picking up stray noise. However, some of my expert readers insist that such noise is well away from the audible spectrum.

If and when you are ready to invest in high-end speaker cables, the criteria listed above can be helpful in narrowing down your prospects to a tiny handful of options. I approached a major cable manufacturer to find one that was both 12-gauge and fireproof. Of the dozens of speaker cables offered by that manufacturer, only a few suited my requirements. When I added heavy insulation to my criteria, only two remained, and I ordered one of each (Monster M1.4s, biwired; and M1.2s).

Another benefit of premium cable is **termination**, or the sealing of bare wire tips inside various kinds of hardware. Copper, the material used to make most cable conductors, corrodes quickly when exposed to oxygen. Anything that can be done to seal it, even a few drops of solder, will extend the life of the cable. That in turn will protect your investment in premium cable and prevent you from having to restrip or replace it. Unterminated generic cable isn't necessarily a bad idea, especially when you're just starting your system, but it's not the best choice for a permanent installation that will go undisturbed for many years.

Cable termination may use hook-like **spade lugs**, plug-in **banana**

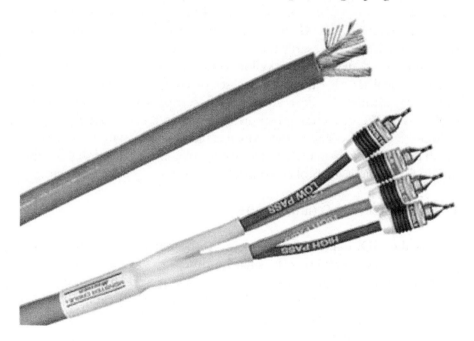

The author's reference speaker cables are Monster 1.2s ($5/foot) and Monster 1.4s ($7.50/foot, shown)—the biwire version. They are 12-gauge in thickness, heavily insulated, and fireproof.

plugs, or narrower **pin plugs**. Spades are the audiophile choice for binding-post terminals because they provide the greatest area of **surface contact** (some people use the phrase **surface geometry**). However, the collared binding posts on most receivers cannot accept spades—in that situation banana plugs are a better option. As a reviewer who constantly swaps speakers in and out, I use bananas and they work well. If your speakers have cheap wire-clip terminals, pin plugs (along with soldered or bare tips) are the only options. Otherwise pins should be avoided because they provide the smallest area of surface contact.

Video interconnects

Let's briefly review what past chapters have said about video interconnects. **HDMI**, **IEEE 1394**, and **DVI** are high-quality digital HD-capable interfaces. **Component video** is analog but still HD. In the not-HD category are **S-video** and **composite video**.

The good news about video interconnects is that most of them come free with other things you buy. A Blu-ray player, if you're lucky, will come with an HDMI or component video cable and some audio cables, though not necessarily the ones you need.

More good news—three yellow-coded composite video interconnects of equal length can pass for a set of component video interconnects. One composite video interconnect can pass for a digital coaxial cable. Just make sure your substitute cables have a characteristic impedance of 75 Ohms. (Some yellow-coded cables are suspiciously slender.)

The bad news is that these free cables may be flimsy enough—with poorly fitting plugs, or broken strands—to require immediate upgrading.

HDMI interconnects (and hype)

As the HDMI interface has come to dominate home theater systems, the quality of HDMI cables has become an issue. Here the marketing magic of the premium cable manufacturers is at its most intense. A lot of readers have asked how much to spend on HDMI cables. My best advice is to use whatever's supplied with your equipment. If you need more, go to oppodigital.com for sturdy, modestly priced HDMI cables. If your video display is UHD or HD, make sure it's a high-speed cable.

While some manufacturers have come up with elaborate hierarchies of HDMI bandwidth, there are only five official classifications. Here they are, straight from HDMI.org: "Standard HDMI Cable supports data rates up to 1080i/60. High Speed HDMI Cable supports data rates beyond 1080p, including Deep Color and all 3D formats of the new 1.4 specification. Standard HDMI Cable with Ethernet includes Ethernet connectivity. High Speed HDMI Cable with Ethernet includes Ethernet connectivity. Automotive HDMI Cable allows the connection of external HDMI-enabled devices to an in-vehicle HDMI device." Data rates are 2.23 Gbps for Standard, 10.2 Gbps for High Speed with HDMI 1.4, and 18 Gbps for HDMI 2.0. The HDMI 2.0 and 2.0a standards do not add any new cable classifications.

The newest versions are Premium High Speed HDMI Cable and Premium High Speed HDMI Cable with Ethernet. HDMI.org says they are UHD capable and support "higher frame rates, HDR, expanded color spaces including BT.2020 colorimetry, 4:4:4 chroma sampling, and they are identified by the Premium HDMI Cable Certification Label." The HDMI Licensing Organization has instituted a voluntary testing program designed to ensure that Premium HDMI cables have an 18 Gbps throughput to support UHD and HDMI 2.0b (and up) at the highest level of quality. Current applications don't require all that speed, but future ones will. There will also be testing for EMI (electromagnetic interference). Certified cables wear the Premium HDMI label with QR code and hologram sticker.

That's all she wrote. Don't believe anything else you read on this subject.

HDMI cables carry great quantities of data. Like any kind of cable, they are subject to signal attenuation, and the longer they run, the more likely attenuation is to affect performance. Credible sources (outside the cable industry) say HDMI cable performance limitations can affect both picture and sound quality over long runs. Video problems may include noise and blocking. Audio problems may include jitter. If you're running HDMI to a ceiling-mounted projector, a long run is unavoidable, and you may run into problems. Here are some solutions.

One solution is to buy HDMI cables with the **Redmere** chipset. All HDMI cables have chips but Redmere chips siphon a small amount of power to enable longer runs over skinnier cables. Note that they are unidirectional, which would disable the Audio Return Channel. Redmere cables go up to 60 feet.

Another solution is to use signal amplifiers, equalizers, or repeaters. Another is to convert HDMI to component video at one end and back to HDMI at the other end. Another is to use component video all the way—it's analog but HD-capable and stands up to long runs. Or try

HDBaseT, which converts HDMI to cheap Cat5 cable and back, and also tolerates long runs. A custom installer can help sort out the options.

Digital audio interconnects (optical or coaxial)

Digital coaxial cables are so named because they are constructed with an outer sheath covering an inner core (a description that, coincidentally, also applies to the otherwise different RF cables that connect cable and other set-top boxes). Any yellow-coded composite video cable with characteristic impedance of 75 Ohms will do the job, in addition to the high-end cables made and sold specifically for digital use.

Digital optical cables send pulses of light through a filament. **Toslink optical cables**, which use a plastic filament, are the most common type. The plug snaps neatly into the jack. Optical connections get dodgy when the cables get kinked (treat them gently) or when jacks are contaminated with dust (keep optical jacks covered, and optical cables plastic-bagged, when not in use). There are also digital optical cables with glass filaments in some high-end two-channel gear.

Audiophiles have long debated whether digital coaxial cables are better than digital optical ones. Manufacturers are concerned enough about the buzz to include both jacks on many products such as receivers and disc players. If there is a difference between optical and coaxial cables, I have never been able to hear it, and I've listened for it through some fairly good equipment. However, I have received some passionate email from readers on this topic, and both camps make good points. Coaxial cables have a higher bandwidth, and are certainly sturdier than optical cables. On the other hand, optical cables are immune from EMI and ground-loop hum. And the debate rages on.

For a USB DAC feeding high-res audio from a computer to an audio system, the right **USB cable** is crucial. I reviewed a DAC that was plagued by dropouts until I replaced the cheap generic cable with a slightly better generic cable with gold tips and a ferrite core filter. There are pricier alternatives I haven't tried; I can't vouch for or against them.

Analog audio interconnects

In the analog sphere, interconnects are more likely to sound different from one another. If you're a high-end audio buff, and you've gone out of your way to buy a high-end preamp-processor or receiver with pure analog signal routing, high-quality analog interconnects are a necessary investment—both for oft-used source components, and especially for the amp/pre-pro interface. Also, if you're using the 5.1- to 7.1-channel analog line outputs of a Blu-ray, DVD-Video, DVD-Audio, or SACD player to feed a receiver or pre-pro, using high-quality cables makes a great deal of sense. For less frequently used components, premium analog interconnects are something you might consider fooling around with *after* you've matched your components, perfected your speaker placement, and paid off your mortgage.

For further reference

An excellent source of unbiased information on the controversial subject of cable is Stephen H. Lampen's *Audio/Video Cable Installer's Pocket Guide* (McGraw-Hill).

Racks, stands, & mounts

More than a mere afterthought, audio/video furniture should be considered another component in the system. It affects how your system looks—to you, family members, and guests—and even how it performs.

Audio/video component racks

Unless you want to buy a new rack every few years, make sure anything you buy can accommodate not only your current system, but anything you may add to it in the future.

Naturally, you don't want to bring anything ugly into your home. Is that "wooden" rack real hardwood or a veneer- or vinyl-covered particle board? There's nothing wrong with the latter. Lots of good entry-level racks are made of **medium-density fiberboard** (**MDF**), an acoustically neutral material that's also used to make most loudspeaker enclosures. But it doesn't hurt to know what you're buying.

Some audiophiles prize steel-frame racks for their high rigidity and vibration control. Shelves may be MDF, metal, or glass. I don't recommend glass racks. They've been known to spontaneously shatter due to temperature changes, even when holding less than their specified weight.

Cable management features make it easier to add new components to your system. That's not to say that the easy way is the best way. Avoid **built-in power strips**. True, they make it easier to neatly arrange power cords, but a cheap power strip is no substitute for a high-quality surge suppressor—and may pose a fire hazard. A **built-in light** can be helpful, though, by illuminating the back-panel area where techies spend hours of our lives.

If you can't remove dust on, under, and around your system, it becomes a dirt magnet. **Casters** are one way to avoid that, though hardcore audiophiles prefer **spikes**.

Speaker stands

Choosing the right speaker stands is a question of performance as well as appearance. Flimsy speaker stands may subtly undermine your system's sound. Stands may have a **resonant frequency** that's audible when the bass guitar or kick drum hits a certain note.

In other words, they have a note of their own, and that note is audible when you rap on them with your knuckles. To prevent that note from affecting the musically significant midrange, the most committed audiophiles will do desperate things, filling **hollow** metal or MDF stands with sand or shot to muffle the stand's resonant frequency. (Sand and cat litter are safe choices. Lead shot may pose a health hazard.)

How the bottom of the stand meets the floor also affects sound quality. You have two choices, **spikes** or **feet**. Spikes couple the stand to the floor. They draw resonance out of the speaker cabinet, producing cleaner sound, and conduct bass through the floor, which can be quite exciting, both for you and your downstairs neighbors. To spare them the movie ballistics, use rubber feet instead.

Speaker mounts

Mounting a speaker is never sonically the best option. For good sound, stands or freestanding speakers are always best. Not all speakers sound good when placed close to the wall—in fact, most don't—but some are actually designed to be used that way. Unlike floorstanding speakers, small satellites, flat speakers, and soundbars are often good candidates for mounting.

There are two main kinds of speaker mounts: keyhole and threaded insert. The **keyhole mount** is simplest. If your speaker has a keyhole on back, just sink a long nail into the wall, preferably into a stud, hang the

186

speaker, and relax. Heavy speakers must be mounted into studs, not merely into drywall.

The best union of speaker and mount comes when speakers are equipped with one (or better yet, two) **threaded inserts**, preferably 1/4-20, the OmniMount standard. Some speaker models come with other sizes of threaded insert such as 4mm/8-32 or 5mm/8-32.

A speaker mount consists of a back plate that mounts to the wall, and a plate or other part that bolts to the speaker, usually separated by a hinge. The more metal parts, the better, although OmniMount does make a successful and popular hybrid with a plastic ball joint that swivels at a variety of angles.

Buying the speaker brand's mount is rarely a bad idea even if it costs a bit more. One popular speaker manufacturer designs all of its mounts to hold seven times the speaker's weight.

Be careful to compare the mount's **rated weight capacity** to the actual weight of each of your speakers, as specified in the speaker manuals. The safety-conscious consumer will buy a mount rated to hold several times each speaker's weight. Another benefit of buying a stronger mount is that the hinge will be stronger and therefore less likely to flop or wobble under the speaker's weight. You won't have to go crazy trying

The OmniMount 20.0 W holds speakers up to 20 pounds. The polymer ball fits into a steel wall-mount bracket and can be rotated both horizontally and vertically.

to aim the speaker in the right direction.

Consider what kind of **wall** you have. It may be drywall, double drywall, plaster, concrete, or brick. A typical suburban home has single or double drywall over wooden studs. One stud is located with a **stud finder** (an inexpensive device available at your local hardware store) and holes are bored to receive the mount's back plate.

Speaker mounts can be invasive, poking holes in walls and their underlying supports. If you're renting your house or apartment, speaker (and TV) mounting may be a bad idea. You may either violate your lease or leave a mess for the people who inhabit the space after you're gone.

TV mounts

TVs neatly mounted to the wall needn't be limited to pricey magazine spreads. Anyone who can knowledgably and precisely penetrate the wall of a house or apartment can mount a TV on wall or ceiling. If you can't do that, find an installer, dealer, or store that can. For heavier sets (like plasmas) and more problematic situations—for instance, you don't want to violate your lease—consider a TV stand instead.

As with any TV installation, **screen size** and **viewing distance** are interrelated. The viewer should be about three screen heights (or more) away from an HDTV.

A TV mount always has at least two parts: a **back plate** that mounts onto the wall, and a second plate, either a **front plate** to hold a flat TV or a **platform** to hold a direct-view TV. What happens in between affects both the mount's overall stability and its ability to move. Some TV mounts are **fixed**, while others **swivel** or **tilt**. That can help with an LCD TV, given its limited viewing angle. Look for **cable management** features. Check **clearances** to ensure that the tilting TV doesn't hit the wall or cabinet doors.

Be absolutely certain that the mount's **rated weight capacity** is enough to safely hold both the TV's weight and, don't forget, that of the mount itself. Get your TV's weight from its instruction manual. The safest mounts are rated to hold several times the weight of your TV (and must live up to that rating!). With an adjustable mount, there is even greater need for the mount to support and manipulate the weight of the TV. Search online for product recalls and safety alerts, either on store sites or on the website of the U.S. Consumer Products Safety Commission: www.cpsc.gov/Recalls.

As with a speaker mount, stop to consider what kind of **wall** you have, and how it may be penetrated, if at all. Mounting kits may allow for various types of wall, but TV mounting isn't a good idea for every wall. Drywall or plaster alone will not safely hold a large TV—you have to **penetrate a stud** behind the wall.

Some sets may be too heavy even for a **single-stud mount**. Then you'll need a **double-stud mount**, using two studs behind the wall. In new construction, it's a good idea to install double studs for future use. **Wooden studs** are stronger than metal ones. Metal studs are not advisable for large TVs. A **recessed mount** needs an additional wall plate for support and must have two inches of clearance all around for ventilation.

The more complex the installation, the better off you'd be to leave it to an experienced and carefully vetted installer. If you need help, **CEDIA** (the **Custom Electronic Design and Installation Association**) knows someone who can do it right. See cedia.org.

TV stands

The right TV stand will safely support your set's weight—or more than its weight, to be certain—and will do so at the proper height for viewing. Consult your TV's manual to find out how much it weighs, and make sure anything you buy can support that weight, and then some.

Safety is an issue, especially in homes with children. A busy toddler loves to yank on cables and could easily pull a poorly situated TV down on its head. This results in tens of thousands of emergency-room visits every year. Your TV stand should be stable and solid, with wide legs or a solid base. The TV should sit low to the ground and near the back of the stand. Attach the set to the wall with safety straps or an L-bracket. Heavy items should go on shelves close to the floor.

Some TV stands have shelves made of glass. It's supposed to be **tempered glass**, which disintegrates into pebbles (instead of angular shards) when broken. Despite the claims of manufacturers, some glass stands have been known to shatter under as little as half of their rated weight limits. Don't buy a glass stand.

For a bedroom TV, assuming it's too large to perch on a dresser, the simplest option is a **pedestal**, perhaps with a **swivel** to allow viewing from different parts of the room. But a TV used with source components needs a **rack** with space for components beneath the set.

The largest-scale option would be an **entertainment center** that holds the TV, components, and assorted media. These hulking things

189

sometimes waste more space than they save, but if you want one, verify that it's a good fit for your TV's height, width, depth, and ventilation needs. Do not enclose a TV with side- or rear-mounted speakers in an entertainment center unless you don't plan to use them at all.

If the system includes a front center speaker, also consider where it will go—is there a shelf for it? Will the entertainment center leave room above or below the TV?

Connecting a Home Theater System

Quick guide

- Pick a hub
- Position components
- Speaker placement
- Subwoofer placement
- Connect video display
- Connect speakers
- Connect picture and sound sources
- Connect network
- Connect power
- Test system, source by source
- Run channel scan
- Arrange lighting
- Adjust picture settings
- Run auto setup and room correction
- Adjust surround levels
- Other surround settings
- Bass management settings
- Adjust subwoofer controls
- Program remote
- Acoustic tweaking
- Mechanical tweaking
- The upgrade itch

Installation guide

A good home theater system isn't built in a day. Setting one up is not something you do after getting home from work. Pick a lazy weekend when you've got time to do things right and attend to the fine points.

Some home theater products (for example, projector mounts or high-end surround processors) are so sophisticated that they require a dealer's help to set up. Those buying high-end gear should use a knowledgable installer to get the performance for which they've paid.

Pick a hub

Every system needs a central switcher. Therefore setting up a home theater system begins with a major strategic decision: Where will its heart be? Home theater systems follow one of four patterns, connecting:

- through a receiver,
- through a preamp-processor
- through a television set, or
- through a soundbar or soundbase

The **audio/video receiver**, or **surround receiver**, is the logical system hub for a practical home theater. While it makes your system a little more complicated than a TV-centric system, a receiver-centric system brings several other significant benefits. First of all, sound quality will be vastly better than anything a TV could provide. You have the opportunity to buy better speakers than those in your TV (and that isn't hard to do). A good receiver will provide enough power to make your five or more speakers sing. And it will throw in plenty of switching to handle your disc player, DVR, streaming box, mobile device, CD jukebox, turntable, or satellite/cable/other boxes—as well as anything you might add in the future. Practical doesn't mean miserly!

Those who want heavier artillery will gravitate to surround separates comprised of a **preamp-processor** and 5.1-channel (or more) **power amplifier**—or possibly even a rack of stereo or mono amps. Separating the amp's hot, power-sucking circuits from the preamp theoretically should let both do their jobs better, resulting in cleaner sound. However, such a sprawling system is not practical for everyone with its extra devices, cabling, and power consumption.

Connecting through a television set is the most convenient option. It's something you might do in a bedroom system, especially one whose dominant use is talking-heads telly. However, even practical home theater requires large enough quantities of picture *and* sound to create a true suspension of disbelief. That total immersion in the story is something the puny sound systems in TVs cannot provide. They can't compete with receiver-based sound systems.

Connecting through a soundbar or soundbase is the newest option. Whether it's advisable depends on the soundbar. I recommend against using a bar as system hub unless it has internal amplification, lossless DTS-HD Master Audio and Dolby TrueHD surround decoding, and HDMI connectivity to accept input from HD video source components and feed output to the TV. HDMI-compliant soundbars are rare but growing. Other bars should be connected through the TV, making the TV the system's a/v switcher. Many soundbars are active, meaning they have internal amps, but most omit video switching, and their surround decoding is for lower-quality formats such as Dolby Digital and DTS. Passive soundbars support only the three front channels and have no internal amps. Treat them as the three front speakers in a receiver-based system, not as a hub.

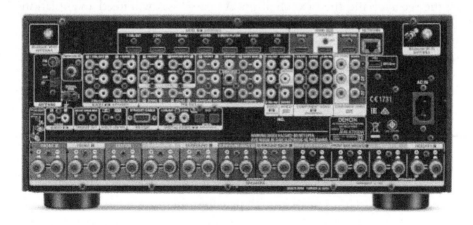

The Denon AVR-X7200WA serves its 9 amp channels—enough for Dolby Atmos 5.1.4—with 11 sets of binding posts, allowing flexibility for multi-room use.

Position components

As you uncrate components, put the cartons into storage, and keep them at least for the warranty period of each product. You'll need them for any components that have to go out for servicing. If space is tight, flatten them and store the foam inserts in trash bags. Create a file for instruction manuals so that you can find them when needed.

Get all components in position before you start wiring them. Don't just stick them anywhere. Stop to think about their ventilation needs. Consider which front-panel controls you'll be using most often. For example, if the disc player's play and eject buttons are below the disc drawer, they'll be harder to reach if the player is too low on the rack.

Place the video display (and seating) just beyond the point where dotted pixels (picture elements) are visible. If you're positioning a wide-screen digital set, the minimum viewing distance is about 1.5+ times the screen diagonal (or search "viewing distance calculator"). As long as you don't see the dots that make up the picture, you're in the ballpark.

Place other components where their interconnects can reach your receiver or preamp-processor. Be sure to allow adequate ventilation for all equipment. Don't block ventilation holes on the top or sides. The best place for a receiver is atop the rack, where it'll have the most breathing room. If something else must go on top—a turntable, for instance—be sure to leave several inches for ventilation between the top of the receiver and the next shelf up. A gear closet will need a ventilation fan to suck out heat and keep components from overheating.

Speaker placement

Place speakers where the instruction manual recommends. (That may sound like a cop-out, but fact is, different speakers are designed to be used in different ways.)

You may wish to experiment with different placements of the oft-neglected front-center speaker. If it is on top of (or above) the TV, try angling it down toward the listening position by placing something under the back of the enclosure. If possible, better to put it below the screen, angled upward. The aim, in either case, is to match tweeter outputs across all three front speakers, so the front soundstage maintains continuity when a sound pans from side to side.

Audiophiles prefer their main front left and right speakers a few feet from the wall because, for acoustic reasons, that's where they usually sound best. In some homes, audiophile-approved speaker placement may intrude into the room. But if you've invested in good speakers, don't shove them up against the wall where they won't sound good.

Imaging—the system's ability to place objects accurately in the soundfield—can be inhibited in home theater systems that place a video display between the front left and right speakers. The problem is **dif-fraction**, or sound waves bouncing off the TV. To minimize acoustic interference from the video display, place the front left/right speakers slightly farther out from the wall than the front center speaker, so that the three front speakers form an arc, rather than a straight line, and are equidistant from the listener.

Another highly effective way to adjust both imaging and tonal balance is to **toe in** the main speakers toward the listening position. This maximizes the proportion of sound heard on-axis. Some speakers are designed to work best when toed in; other are designed to sit parallel to the side walls. Toeing in generally produces more accurate imaging, along with stronger high frequencies. If you find the highs fatiguing, start reducing toe-in till you find the optimum tradeoff between accurate imaging and pleasing tonal balance. Another option is to put a wedge under the front of the speakers, or use lower spikes in back. That will not only reduce fatiguing highs, but will also minimize standing waves, because the sound will hit the back wall at a different angle.

If you listen often in stereo, using full-range front left/right speakers, positioning of those main speakers is critical. Bass will always be stronger when speakers are placed nearer the floor, walls, and corners. That's because such positioning will reinforce the tendency of bass waves of certain frequencies to bounce between parallel walls, a phenomenon called **standing waves**. Also, in different areas of the room, bass waves will either **reinforce** or **cancel** each other, something you can easily hear by putting on a low-frequency test tone and walking around. Systems heavily dependent on a single subwoofer (as opposed to two subs or full-range speakers) are especially subject to these effects. But don't assume you should always place speakers and subs to maximize bass. Too much bass can lead to a muddy, incoherent, and ultimately unsatisfying presentation. Instead, look for the right tradeoff between clarity and power.

Speakers that mate with flat-panel displays are increasingly popular. They're designed to mount on the wall, along with the display, so put them there. However, be sure to keep them all on the same level—above or (more likely) below the display. Do not put the left and right speakers alongside the screen unless you have only a stereo pair.

Surround speakers (as they are called in a 5.1-channel system) or side-surround speakers (as they are called in 6.1- and 7.1-channel systems) belong on the side walls, toward the rear of the room, while back-surrounds should be against the back wall. Their effects are supposed to be diffuse, so on-wall or in-wall mounting would be more acceptable. If you're into surround for music listening, back-wall placement is the rule. Experiment, if you like, but in a permanent 5.1-channel speaker installation, stick with side-wall placement. Later you'll experiment with speaker placement to adjust imaging, bass, etc.

A common flaw of many surround systems is that the rear speakers are aimed too directly at the listener. **Bipole/dipole** models avoid this by firing to the front and rear of the room, leaving the listener in the **null** area (the place with least sonic energy). To achieve the same effect with non-bipole/dipole rear speakers in a non-Atmos system, try aiming them at the ceiling. This works especially well if your system requires placement of rear speakers near the listener—say, on either side of a sofa. You can even do it with wall-mounted speakers. When the spray of sound hits the ceiling, it will be dispersed evenly and diffusely over the seating area. You'll need to set the receiver's surround levels a little higher to compensate for the reduction in direct-radiated sound.

If you are adding **Dolby Atmos-enabled speakers**—the kind with upward-firing drivers at the top for height effects—Dolby recommends that you avoid placing them higher than one-half of wall height. They should be at least three, and ideally more than five, feet from the listener. Add-on Atmos modules should go on top of the front and side-surround speakers or within three feet of them.

If you are adding two separate ceiling speakers for Atmos, they belong slightly in front of the listening position. If you are adding four separate in-ceiling speakers for Atmos, place the front pair in front of the listening position and the rear pair behind the seating position.

If you are adding height and/or width speakers to your system, Audyssey recommends placing front left and right speakers 30 degrees off center, height speakers 45 degrees off center and another 45 degrees upward, width speakers 60 degrees off center, and side-surrounds 120 degrees off center.

Subwoofer placement

Bass response from a subwoofer will be most powerful in corners because they emphasize bass. However, corner placement is likely to give bass a booming one-note quality. It's better to keep your sub somewhere on the front wall, between the left and right speakers, or on the side walls, between the front and rear speakers. The sub in a compact sat/sub system should go as near the front-center speaker as possible because it has to deliver more upper bass. In that situation, you'll get a better vocal blend if sub and center are close together.

Here are some tricks that'll help you find the ideal subwoofer placement: Put the subwoofer in your listening position and walk around the room till you find the spot where the bass sounds best. When you reverse the positions of yourself and the sub, the same acoustic conditions will apply, and the subwoofer will sound as it did when you first stepped into the magic spot. Another technique is to temporarily wire the main speakers out of phase (red to black, black to red) and adjust the sub's phase control till bass is weakest. Correct the speaker wiring to make it in-phase (red to red, black to black). This should put speakers and sub in phase with one another.

For a firm theoretical grounding in the science of acoustics, I recommend *Sound Reproduction: Loudspeakers and Rooms* by Floyd E. Toole (Focal Press) as well as *The Handbook of Acoustics* (4th Edition, McGraw-Hill) by F. Alton Everest.

Connect video display

If your system hub is a receiver, pre-pro, or HDMI-compatible soundbar, connect its monitor outputs to the video display. HDMI is the best choice in most systems. Newer receivers convert incoming video signals from any input and send them through the HDMI outputs. In that case just make a single HDMI connection to the video display. Older receivers may not convert video signals from one format (component video, etc.) to another. These video streams are sometimes kept separate within older receivers.

If you're connecting a front-projection display, be warned that not all interfaces work well with long cable runs. Best in this regard is the HDBaseT standard, which travels up to 328 feet. Also good are 1394 and component video, both which can survive runs of up to 300 feet. An RGBHV connection can travel up to 100 feet, an HDMI connection up to 75 feet, and a DVI connection up to 50 feet. In some cases signal convertors can extend these limits. Consult an installer.

Connect speakers

Premium speaker cable can be a good investment (if you don't overspend). But before you lay out big money for premium cable, start your system with generic wire ($15 for 50 feet). Using cheap cable first will give you a chance to perfect your cable runs, experiment with different configurations (such as biamping and biwiring), and avoid subsequent, expensive mistakes. Don't worry about pricey premium cable till you've broken in all the other components. Generic cable is, among other benefits, both inexpensive and fairly neutral. Do premium speaker cables make a difference? I believe they do, but you've got to graduate from high school before you go to college, so to speak.

With standard binding posts, use 16- to 12-gauge cable (thicker cable for longer runs) to connect your speakers. Note that as the gauge (AWG) number goes down, thickness increases. Measure each cable run from amp to floor to speaker and allow a few feet of slack. Cables for the three front speakers should match one another within a fraction of an inch. The same goes for each pair of surround or other speakers. Strip off a half-inch of insulation using a $10 wire stripper from the hardware store. It resembles a pair of pliers with different-sized notches for different cable thicknesses. Twist the strands of copper together.

Speaker-cable tips can be **terminated** (sealed) if desired. **Spade lugs** (Y-shaped hooks that fit around the binding post) are reputed to provide the highest-quality connections because they have the largest area of surface contact. However, spades do not fit the collared binding posts on receivers. To my ears, flexing **banana plugs** sound just as good, and they do fit receivers. Plugs are also favored by some speaker designers and reviewers because they make it easy to swap speakers and cables. Banana-plug adapters, sometimes called **double banana plugs**, can accept bare wire. If you want to use premium cable with sealed tips to feed speakers or a receiver with wire-clip terminals, get cables terminated with **pins**. This provides a very small area of surface contact, so I don't recommend it—bare wire is better in this instance.

Speaker connections should always follow the standard color coding found on speaker terminals at both ends. Red always goes to red and black always goes to black. The red terminals are the **positive** (+) or **hot** connection, and the black terminals are the **negative** (-) or **ground** connection. More simply put, current goes in one side and comes out the other. This helpful color coding is also included on better speaker cables. If you're using generic cord, with two leads grafted together, you

198

can figure out which end goes to which end by looking for a stripe, or a ridge, on one side.

If you mismatch a speaker connection, the affected speaker will run **out of phase**. In other words, the drivers will move in when they should be moving out, and vice versa. This can result in a bass-cancellation effect audible as hollow, disembodied, unsatisfying bass response. If your first impression of your system is that something is askew, disorienting, weird—then check your speaker connections immediately. Some receiver auto setup programs will identify this error and urge you to correct it.

Biamplification allows a speaker with more than one set of binding posts to receive more than one channel of amplification. All other things being equal (such as the quality and quantity of amplification), biamplifying your system will definitely make it play louder and cleaner. However, biamping can cause problems when attempted with mismatched amps. Either the amps must be of the same design, or one must have a gain (volume) control so you can match them with a 1 kHz test tone and a voltmeter. Many 7.1-channel receivers make biamping easy in 5.1 by repurposing the extra pair of amp channels. Otherwise, there are easier ways to make your system more dynamic—by upgrading the amplifier, or moving the speakers. Biamping should be thought of as a last resort for huge rooms that need more than one muscle amp.

Biwiring duplicates runs of speaker cable, without additional amplification, using two sets of binding posts at the speaker end and one at the amp end. This gives each driver a separate path to the amp, so the power-hungry woofer can no longer collapse the soundstage by drawing power away from the tweeter. The effects of biwiring are more subtle than the effects of biamping—in a budget system, don't bother. Biwiring is easier with premium cable—just tell the dealer that's what you want. But also keep in mind that biwiring and biamping double the amount of cable needed and therefore double the cost. Experiment with generic cable before investing in the fancy stuff.

Your receiver may have an **Ohm switch** on the back panel. It's designed to limit power output to prevent overheating. If your speakers have a low sensitivity rating (see "Surround speakers/Interpreting speaker specs") it might be wise to use the 8 Ohm setting to protect the receiver—actually, it might be an even better idea to arrive at a more appropriate pairing of speakers and receiver. Otherwise use the 4 or 6 Ohm setting to maximize power output. If your receiver is running not just warm but extremely hot, use the 8 Ohm setting, especially if the receiver shuts down more than once.

For the subwoofer, use either **line level** (low level) or **speaker level** (high level) interconnects. In other words, you may connect it using the same low-level interconnects you use to link other components to the system, or you may use speaker cables, which carry more current. The line-level (LFE or sub-out) connection from receiver to sub is usually best because it allows speaker-level connections (which handle everything above the sub crossover) to go directly from receiver to speakers. The speaker-level connection is occasionally a better choice for the smallest satellite/subwoofer sets because they are designed to function with a fairly high crossover from about 120 to 180 Hertz using the speaker-level crossovers built into their subs.

Most subs come with both a low-pass filter and a high-pass filter. The **low-pass filter** is variable, and adjustable via the sub's crossover control. It ensures that the sub reproduces only sounds that fall below the crossover frequency. If the crossover control is set too high, you'll start hearing voices booming out of the sub. A receiver or preamp-processor contains a low-pass filter of its own which duplicates the one in the sub. If you were discerning enough to buy a sub that can switch out its own low-pass filter in favor of the receiver's, take advantage of that bypass mode. Otherwise your system's low-bass frequencies may suffer from having to pass through two filters, a process known as **cascading**. This is discussed in greater detail in the upcoming section on how to "Adjust subwoofer controls."

The **high-pass** filter is usually fixed, and used with a speaker-level connection. The receiver sends a full-range signal to the sub, where the high-pass filter strips out the lows, and passes the rest of the signal to the satellites. In rare cases, the high-pass filter may be used with a line-level connection. Either way, this may result in more dynamic sound, with less duplication of bass response—but at the expense of signal purity, because the signal has to travel farther, and through a filter.

If your receiver has more than five channels, you'll have to configure your receiver for use of the extra channels. We'll cover that later.

Connect picture and sound sources

It's time to link your system hub to signal sources such as Blu-ray, DVR, cable, etc. Here are a few important rules for interconnecting the components in your home theater system:

- Connect input to output, output to input.
- With older receiver, use same connection throughout signal chain.
- Connect video and audio to the same-named input.

First and foremost, observe the golden rule for any form of component interconnection: in-to-out, out-to-in. Do not plug an input into another input or an output into another output.

Most receivers and preamp-processors provide **video upconversion** capability to translate one video mode to another, usually routing all incoming signals (regardless of input type) through the HDMI or component video outputs, enabling a single connection to the TV. However, older receivers require the video signal to travel over the same path from source component to receiver to TV.

Finally, for a non-HDMI component requiring separate video and audio connections, group them under the same input name. If you connect your DVD player's component video output to "DVD," don't connect its audio outputs to "DVR." If you must deviate from the receiver's naming scheme, most receivers let you reassign or rename inputs. Digital audio connections, like their analog counterparts, will be assigned to specific inputs. Whether a digital connection can override an analog connection made to the same input will depend on the design of your receiver or pre-pro. It may be necessary to go into the receiver's input menu and assign an HDMI, coaxial digital, optical digital, or analog connection to each input.

Now, here's an overview of ways to make video connections:

- HDMI
- DVI-HDCP
- IEEE 1394-DTCP (DTV Link)

If possible, use these digital video interfaces—they afford the highest-quality connections. The best of the best is **HDMI 2.0a**, which handles all current video and surround standards at top resolution and highest quality. The **1394** interface is suitable for recording and networking; the HDMI and **DVI** interfaces are not.

In pre-HDMI components, the highest-quality video comes from analog connections that do the most to separate the video signal into its primary elements. It's best to use the highest-quality video connection supported by your video display. The best analog video connections, most of them found on receivers, are in descending order:

- component video
- S-video
- composite video
- RF coaxial

Component video is the highest-quality *analog* video connection found on a receiver, and the only HD-capable analog connection. It may be labeled in any of several ways including Y Pb Pr, Y Cb Cr, or YUV. Component video cable is often color-coded red, green, and blue to help avoid confusion. In any event, component video splits the video signal into three parts, including a brightness signal and two color-related signals. Three composite video cables with the correct characteristic impedance (75 Ohms) may substitute for a set of component video cables. When HDMI and DVI are not available, component video is the best way to link an HD source to an HDTV, though it's being phased out.

S-video uses a multi-pin cable whose delicate pins separate video into brightness and color, thus preventing visible problems such as cross-color interference and dot crawl. S-video is the best way to connect S-VHS VCRs and older equipment without HDMI or component video. But it is not HD-capable and is disappearing from receivers.

Composite video combines all the video components into one cable, color-coded yellow. It's useful mainly for connecting an iPod dock, VHS VCR, or other legacy source, and is not HD-capable.

RF (radio frequency) coaxial cable carries multiple channels of picture and sound. It's used to connect an antenna or cable feed to your system, or to feed an ancient TV set (via channels 3 or 4) that has only an antenna-in and no direct video (HDMI, etc.) connection.

As for audio connections, HDMI is the best choice, digital coaxial/optical second best, and analog in last place—partly because most receivers convert all analog inputs to digital. Using direct digital connections helps avoid that unnecessary analog-to-digital stage.

Finally, here's a rundown on how to connect each source component in your system.

Blu-ray and DVD players: Use the highest-quality video connection supported by your video display. Today that means HDMI 2.0a for HDR-savvy UHD. Tweakers seeking extra credit may do some experimenting to determine which has best video processing: the disc player or other source component, the receiver, or the display. If you're using a UHD or 1080p display, start by having the source component output UHD or 1080p. You can then set the receiver to process or simply pass through the signal. Then you can experiment with settings in the display.

On the audio side, newer Blu-ray players use HDMI to output a bit-stream to feed receivers with on-board decoding of the new lossless and other surround codecs. In that case, use HDMI. Dolby TrueHD, Dolby Digital Plus, DTS-HD High Resolution Audio, and DTS Encore are supported by HDMI 1.1 and 1.2 (and up) if your receiver has one of those inputs. For DTS-HD Master Audio your receiver will need HDMI 1.3 or up. If your older receiver does not have onboard decoding for the new codecs, but does accept high-res PCM, you can use the player's de-coders and set it to output PCM with no loss in quality. If your receiver has neither onboard TrueHD, et al decoding or high-res PCM input, use the multichannel analog interface if it's available on the player.

To obtain old-school Dolby Digital and DTS surround signals from either Blu-ray or DVD, HDMI suffices. For pre-HDMI receivers, con-nect one of the player's digital outputs to one of the receiver's digital inputs. For an older DVD-Audio and/or SACD player that has 5.1-channel analog line outputs and no HDMI, use the analog connection. It may degrade the signal slightly, so a digital connection is preferable.

HD-DVR: Using the DVR supplied by your cable or satellite pro-vider is the best course. Use the highest-quality video output, generally HDMI. Otherwise, a digital video recorder may require a direct link from your broadcast/cable/satellite set-top box using the IEEE 1394-DTCP interface—unless it has ATSC (broadcast) or QAM (cable) tun-ers. Then it requires an RF input. CableCARD-equipped gear requires an RF-in plus a CableCARD from your local operator.

HDTV set-top box (broadcast, satellite, or cable): These devic-es receive RF inputs. It can be convenient to connect the box to both your receiver, for movies and music, and directly to your TV, bypassing the audio system for talking heads. HDMI is the best connection, but for that secondary connection, component video and audio (digital or analog) will do. If your non-HDMI satellite receiver supports Dolby Digital surround sound, connect the Dolby Digital output to a digital coaxial or optical input on the receiver. A phone connection may be re-quired so that the satellite operator may communicate with your system for billing purposes during off-peak hours (usually late at night).

Antenna: Use an RF splitter to divide the signal between the RF inputs of the DTV/set-top box and DVR. That will enable you to rec-ord one channel while watching another, or to use single-tuner picture-in-picture to watch two channels at once. On a more practical level, it will also allow you to watch the news using the TV's built-in speaker(s) and without having to turn on the receiver. Don't buy an eight-way split-ter if you're just splitting the signal two ways—each extra splitter acts as a signal-polluting broadcast antenna.

Smart TV: Requires a broadband connection, wired or wireless, to show off its smarts.

Media server or streaming device: Use the HDMI for highest-quality audio/video connection or optical/coaxial digital outputs for the highest-quality audio connection. Your device also requires a broadband network connection.

iPhone/iPad/iPod: To connect an iOS device to your system, you can use wireless or wired connections for audio and (if desired) video. Apple's **AirPlay** streams both video and audio. It is built into some re-ceivers; others can be retrofitted with an AirPlay Express. It requires a network connection. A wireless audio-only alternative to AirPlay is Blue-tooth, which is built into iPhones, iPads, and newer iPod touches and nanos. Receivers may have AirPlay, Bluetooth, both, or neither, with or without extra-cost adapters—know what you're buying. The wired alter-native would be an **iOS-compliant USB** jack on the receiver, which handles audio only. Note that not all USB jacks on receivers speak iOS. For video as well as audio, use an iOS dock with a composite video out-put, or an iOS-to-HDMI adapter. Some iOS docks use proprietary con-nections; others use standard a/v jacks to interface with any receiver.

Android phone or tablet: For Android devices, there are various wireless protocols—for example, the up-and-coming **Miracast**. Some handle video and audio, others audio only. For a wired a/v connection, there's a variation of HDMI called **MHL (Mobile High-Definition Link)**. The MHL standard is plug-agnostic but usually has HDMI at the TV end and five-pin mini-USB at the other. Some Samsung phones have an 11-pin plug. MHL usually charges the device, so loss of power won't interrupt a movie. These connections must be supported by both your phone and tablet and your receiver.

Other phones, tablets, and music players: To plug in any brand of portable audio player, phone, or tablet, buy an adapter that has two RCA-type plugs at one end (for your receiver) and a 1/8-inch (3.5 mm) stereo mini-plug at the other end (for your device). Plug it into an un-used stereo analog line input in back of the receiver. Before you buy the adapter, be warned that some phones use a smaller 2.5 mm mini-plug. You can also listen to MP3s through a home theater system using a BD/DVD/CD player that accepts MP3s burned onto CD-Rs.

Computer: Your PC or Mac makes a lousy signal source because its relentless multitasking causes timing errors in the bitstream. In addition, its built-in soundcard is probably crude. For better-sounding computer audio, add an outboard USB DAC (digital-to-analog convertor). USB goes in one side and (usually) analog audio comes out the other. Most receivers have USB inputs but only a select few of those accept direct computer input. Most USB DACs support USB 1 mode and bitrates up to 96 kHz out of the box. Windows users, please note that your computer will need a driver to handle audio files above the 96 kHz bitrate in USB 2 mode. See USB DAC instructions.

DVD-Audio/SACD: These high-resolution audio formats are best connected through HDMI 1.1 and up (DVD-A), HDMI 1.2 and up (SACD), or their 5.1-channel analog line outputs. Another alternative is to have the disc player output high-res PCM which can function with most HDMI-equipped receivers now on the market. Older products from Denon, Meridian, and Pioneer have proprietary digital interfaces that work as well as HDMI. But ordinary optical and coaxial digital outputs, sadly, do not support full resolution.

CD: Unless you're certain that your receiver is designed to process analog signals in the analog domain (or has an analog-direct mode), it's safe to assume that it will convert all analog inputs to digital. So no matter how good your CD player's digital-to-analog conversion is, you'll never get the benefit of it—might as well bypass the receiver's A/D stage by using a direct digital connection, coaxial or optical. The D/A converter you hear will be the receiver's. On the other hand, if you chose your CD player partly on the basis of its D/A performance, and your receiver has a pure analog mode, feel free to experiment to see which connection (digital or analog) sounds better. The result is not easily predictable. Trust your ears. By the way, using a digital connection will also make it easier to record from CD with a CD-R/-RW deck.

Cassette: How odd that the analog audiocassette is making a comeback. Use the analog tape loop jacks to connect the recorder's outputs to the receiver's inputs and the recorder's inputs to the receiver's outputs. Otherwise, you're free to treat the cassette deck as a playback-only source and it'll plug into any pair of stereo analog line audio inputs.

Phono: Not just any old analog input will serve a turntable. Why? Because phono cartridges produce a tiny voltage that's not nearly as great as a standard line output. You'll need a dedicated phono input, something found on some but not all receivers, to amplify that delicate signal and correctly process it using a standard equalization curve. If your receiver does have a phono input, unless specified otherwise, it will

205

be compatible with **moving-magnet** phono cartridges, the more common and less costly kind found on affordable turntables (and favored by the hiphop community for better bass and durability under the stress of "scratching"). Very few receivers have phono inputs switchable between moving magnet and **moving-coil** cartridges, which may produce more extended highs, and generate even tinier voltages. If you have a large library of LPs and/or 45s, and your receiver has no phono input (or the wrong kind), don't lose heart—just add a separate **phono preamp** to your system, plugging it into any available stereo analog line input.

AM/FM: Connect the supplied antennas per the instruction manual. Most receivers use threaded 75 Ohm RF-type terminals. In areas with weak FM reception, you may want to (further) split your TV antenna feed and use it for radio. Do TV antennas pick up FM radio? They sure do—all FM stations lie between channels 6 and 7 in the VHF-TV band. However, be warned that some TV antennas (and cable systems) use **FM traps** to strip out the FM signals, because they may interfere with VHF-TV channels.

XM or Sirius satellite radio: Some receivers have inputs for next-generation satellite radio, either XM, Sirius, or both. You'll need to buy an external antenna for about 20 bucks and activate an account. Then you can select satellite radio as a source and listen to hundreds of channels. SiriusXM also comes in an online streaming version.

For more on inputs and outputs, see the "Connection glossary."

Connect network

A TV, receiver, Blu-ray player (with BD-Live), streamer, or other component with network-enabled features will need a wired or wi-fi connection to your broadband router. Most products with network features come with ethernet jacks, and some support wi-fi, sometimes with an optional adapter. Connect component to router if you want to access video streaming, internet radio, or other media via DLNA or Apple Air-Play. For DLNA, you need to get the PC to approve the receiver as a network device. In Windows 7 and up, use Control Panel/Sound. In Windows XP, use the Windows Media Player (Tools, Options, Library). For AirPlay you may need to activate Network Standby in the receiver's network menu. If your router uses encryption, you may need to key in its password for each device. Bluetooth does not need a home network connection because it operates directly from device to device. But it does require the two devices to participate in a pairing ceremony.

Connect power

Don't connect power cords until you've done all other connections. Avoid running power cords in parallel with video or audio cables—the power cords will radiate electromagnetic noise into the interconnects.

It may be tempting to connect all power cords to the same power strip or surge suppressor. However, if the power accessory is not of high quality, it can degrade performance. Strongly consider adding a good **power-line conditioner** to your system. A top product will protect components from surges, isolate picture and sound sources from one another's signal-polluting influence, and tell you what's going on with the power line (whether it's properly polarized and grounded).

If you're getting by without a power-line conditioner, at least refrain from plugging a receiver or power amp into anything but the wall, so it can draw all the current it needs. Use your receiver's **unswitched power outlets** to connect components that need a constant supply of juice (DVR, satellite receiver, etc.). The **switched power outlets** can be used for the other components. Note that most components have power on/standby modes rather than a simple on/off arrangement—don't plug these into a switched outlet. They need power to retain settings.

Try to avoid connecting components to a power line that's shared by an air conditioner. As the AC's compressor switches itself on and off, surges or dips may occur in the power line, which could inhibit the performance of your components and possibly even damage them.

Test system, source by source

Now the fun begins—sort of. Turn on the receiver, set it to the AM/FM tuner, and dial in your favorite station. It's better to start with the tuner than with, say, the disc player because if there's a problem, you'll know it's not the source component. Then work your way through all the other components. Make sure a/v sources produce a picture and fill all appropriate channels with sound.

As you do this, keep in mind what you connected to what. Some of your video sources may have been connected through the receiver, others directly to the display. You may have to switch video and audio separately. Later you'll automate that double round of switching by programming your remote control with macro commands. But don't worry about that right now.

Don't be upset if you don't get everything right on the first try (I rarely do). Deal with arising problems by checking connections. If the physical connections are in place, start going through the menus of the receiver or pre-pro as well as other menu-driven components (TV, disc player, etc.). To get everything to work you may need to use the menus to activate, reassign, or rename certain inputs or outputs.

If you get no sound at all from the main speakers, don't panic. A receiver that accommodates a second pair of speakers (over and above the usual 5.1-channel surround array) may have an A/B speaker switch, and you may need to activate it to get any sound at all. In addition, some 7.1-channel receivers allow the back-surround channels to do double duty for height or width speakers, second-zone use, or biamping of the front channels. You may have to switch them from one use to another. If that's not the problem, start looking for loose or missed connections.

For more snares and solutions, see "Connecting a Home Theater System/Problem solving."

Run channel scan

Now that your source components are running—but before you start dealing with picture and sound quality issues—this would be a good time to run the **channel scan** for your TV and any other source components with tuners (DVRs, set top boxes). Definitely do this if you're using an antenna or a CableCARD-equipped TV. If you are a cable subscriber, but receive only the cheapest broadcast-channels-only service, you might want to try this with the cable connected directly to the set—if your TV has a QAM tuner, it may handle unencrypted channels. If you're using a cable or satellite box, you don't need to do this.

First, make sure your antenna or cable feed is connected. You may be asked to specify whether the signal source is an antenna or cable feed, and whether to skip unused or unavailable channels. Start the scan and the device will detect available digital and (where they still exist) analog channels. There may be a manual option but auto is usually easier.

Be warned that some stations fail to transmit accurate channel-associated data. This may cause certain channels to display inaccurate information (or no picture at all). There's not much you can do about it but complain to the station.

When the channel scan is done, you should be able to grab the remote and flip through all available channels. This would be a good time to rest from the rigors of system setup, kick back, and relax for awhile!

Arrange lighting

Depending on the kind of video display in your system, lighting will have to be adjusted.

An LCD TV, thanks to its adjustable backlight, can more easily survive in a well-lit room than a plasma or (especially) projection set. However, it will look its best in a room with low light. Too much light forces you to raise contrast and brightness to levels that shorten the display's lifespan. Too little light forces the eye to focus on both a bright picture and a dark background. That is likely to lead to eyestrain and headaches.

The light should not be allowed to reflect off the screen. So the best place for lighting is behind the set. And placing a shaded low-wattage bulb (25-40 watt incandescent or 7-10 watt CFL/LED) behind the screen will bias the optic nerve, making movie-length sessions more comfortable. Yes, you can actually tweak your own eyes for better performance! For extra credit, raid a well-stocked lighting store for bulbs with a color temperature of 6500 Kelvins (the packaging may specify it). The CinemaQuest Ideal-Lume is designed especially for this purpose.

For wall-mounted sets, a few well-aimed ceiling spots may help. Again, make sure other room lighting does not reflect off the screen

Projection sets are another story entirely. Because they don't produce as bright a picture as flat-panel TVs, they work better in a darkened room. Consider blackout curtains or blinds, especially for daytime viewing. The Eclipse Thermaweave is well made and does a fine job. For best results, choose a dark color.

If you've invested big bucks for a front-projection display for use *in a dedicated home theater room*, paint the walls black or dark grey (avoiding glossy finishes) to give the projector and screen the best possible chance to work their magic without light reflecting from the walls and ceiling. However, be warned that dark walls make the room impossible to use for any other purpose—they suck up so much light as to make the room dim and dismal. In a multi-purpose room, consider darkening just the wall with the screen.

Get a handle on room lighting before adjusting the video display.

Adjust picture settings

Here's a quick guide on how to adjust your video display for best performance—and longest life.

First, realize that your home is not a showroom. Once you get your set home, it won't be competing with a wall of other screens. So there's no reason to stick with factory settings designed to help a set compete under bright showroom lights. In fact, factory settings typically over-drive the set and make it look awful.

Start by setting the display for its **movie** or **cinema mode**. This will immediately get most settings close to ideal. When you're playing games, you may wish to switch to the game mode to avoid video-response lag. However, the game mode is awful for any other purpose—make sure to switch back to movie mode for movies or TV.

For manual adjustment, begin by backing off the **contrast control** (or **white level**). It adjusts the balance of relative black between the brightest and darkest areas of the picture. Adjusting it either makes light areas lighter, or moves everything toward a uniform shade of grey. You're likely to find the correct setting below the factory preset (which may be cranked all the way up). The best setting might even be below the manufacturer's center position, though TVs vary.

Brightness (or **black level**) adjusts the amount of total black in all areas of the picture, light or dark. Too little brightness (or too much black level) brings a picture that is too dark, and starved of shadow de-tail, as objects in darker areas disappear into black. Too much brightness washes out the picture. A correctly adjusted picture on a good set has a deep black—as close to the absence of light output as you can get. True black is more likely with OLED, plasma, and tube-based displays than with LCD and microdisplay technologies, though HDR is changing that.

The **color** control adjusts the overall **saturation** level (intensity) of each of video's three primary colors (red, green, and blue). Again, the factory setting may be too high, or may skew certain colors.

The **hue** or **tint** control (the technical term is **color phase**) changes the balance between red and green. Its primary use is to adjust fleshtones, among the most difficult colors to reproduce accurately. However, fleshtones vary, and to complicate matters, some sets come with fleshtone adjustments that make Caucasian people look right but skew all other colors. So also give some attention to the green playing field in a sporting event (preferably natural turf).

Finally, the **sharpness** (or **detail**) control affects the definition of edges and fine patterns within objects. Most people set it too high in the mistaken belief that they're seeing more detail; what's actually happening is that they're seeing more noise. The correct setting is likely to be from 0-30 percent for Blu-ray/DVD or satellite reception, since these sources produce a very clean picture, and slightly lower for analog video sources.

Some sets can memorize different picture settings for each input.

The second best way to fine-tune your picture settings is with a Blu-ray test disc such as *Digital Video Essentials: HD Basics* (Joe Kane Productions) or the *Spears & Munsil HD Benchmark*. These are useful tools, both for setting up a new video display or tuning up an old one.

But the videophile who wants everything just right will bring in a technician licensed by the **Imaging Science Foundation**. ISF techs calibrate your set using professional test equipment, burrow into control menus not accessible to the user, and make fine-tuning adjustments beyond the skill of an average person. The service may be included in the cost of a high-end video display or you may bring in the technician on your own. The cost is a few hundred dollars—but if you've spent thousands for a high-end video display, it's well worth it. Even moderately priced TV sets can benefit from ISF calibration. It can make the difference between a mediocre picture and a great one. For more information about ISF, visit imagingscience.com.

Run auto setup and room correction

Most receivers can set themselves up automatically and implement room correction to fix acoustic problems. Unless you're an advanced user, you'll want to run the auto setup program. If you don't like the effects of room correction, you can always turn it off later.

Before you run auto setup, make sure that your speakers are in *exactly* in the right places. Silence the room—close the windows and turn off all air conditioners and fans. Place a tripod in the prime listening position, attach to it the small mic that came with the receiver at ear level, turn on the receiver, and plug in the mic. If the receiver doesn't automatically start the auto setup program, start it from the control menu.

The program may ask you some preliminary questions about how many speakers you have and where they are placed. In a seven-channel receiver, the sixth and seventh channels may be configured for back-surround, height, or width speakers. If your speaker setup is basic 5.1, you can also use the extra channels to biamplify the front left and right speakers. Or they can be adapted to multi-zone use. Receivers with more than seven channels allow more of the above options. You may also need to set the sub's volume knob before auto setup begins. Turn it one-third to halfway up. You can always change it later.

As the program runs, your speakers will emit test tones. Be sure not to stand between the mic and the speakers or make any noise—that will throw off the measurements. The test tones may be loud, so either use hearing protection or, better yet, leave the room.

Some auto setups take measurements from more than one listening position to provide better correction outside the traditional sweet spot. You don't have to run the maximum number, but it's a good idea to cover all the main listening positions. If the audience rarely spills off the sofa, take three measurements on the sofa and call it a day.

When the measurements are done, the program may take a few minutes to calculate the right settings. Then it offers you a look at the settings. Make sure the speaker and sub distances look right. If the speakers are detected as "large" or "full range," and you'd rather have the sub handle frequencies below 80 Hz—as recommended by THX—then set the speakers to "small" and select the crossover. Small satellite speakers need higher crossovers, maybe 120 Hz. See their manuals.

If you later move your speakers, even by small amounts, even just a single speaker, run auto setup again. Without correct calibration, a surround sound system can't deliver the enveloping feeling that suspends disbelief and pulls you into the story. Instead, by calling attention to itself, it just competes with the movie. We master technology so that art can take precedence over technology.

Adjust surround levels

As an alternative to auto setup, some surround buffs prefer to fine-tune channel levels manually using a sound pressure level meter. You'll be using it to calibrate the sound levels of each speaker to ensure a uniform surround soundfield. It should be possible to do all this using the remote control from the center listening position.

These instructions assume a reference level of 85 dB. Some home surround gear may include an offset of 10 dB, so it would be correct to calibrate for 75 dB. Check with the manufacturer.

- Turn on meter to BATT setting to check battery.
- Set range control to 80.
- Set weighting curve to C.
- Set response to SLOW.
- Put receiver into surround mode.
- Activate receiver's test tone.
- Adjust each channel 5 dB above 80 within 1/2 to 1 dB.

Turn on the meter. Check its battery strength to make sure it will provide accurate readings. Then set its **range** control to the **reference level** of 80 decibels. Set its **weighting curve** to C. (C-weighting covers all frequencies. The alternative setting, A-weighting, rolls off steeply in the midrange and bass.) Finally, set the meter's **response** to SLOW to stop the needle from jumping around too much.

Put the receiver (or pre-pro) into one of the surround modes and activate the test signal. It's **pink noise**, spanning a wide range of frequencies, and sounds like a continuous storm of static. The signal may jump from channel to channel, or it may stop on each channel—some receivers will allow you to choose the mode. You're aiming for a volume level of 85 dB, or 5 dB above 0 when the meter is set at 80. (You could also read it as 5 dB *below* 0 when the meter is set at 90.) Try to get each channel within 1 dB of 85 (some processors allow increments as small as 1/2 dB). This method works for THX and other gear where the master volume control does not affect the level of the test tones. However, on most receivers and pre-pros, the master volume control does affect the volume level of the test tones. If that's the case, set the front left channel

The classic RadioShack sound pressure level meter uses a needle indicator, though the modern versions have digital displays. The SPL meter is an indispensable home theater system accessory for those who want to doublecheck auto-setup volume levels.

at 0 dB in the setup menu, and turn up the master volume until the meter reads 85 dB. Use that as the benchmark to balance all other channels.

Note to users of THX-certified home equipment: The reference level may be offset by 10 dB to reduce wear and tear on the ears. In that case, you'd set 75 dB on the meter to obtain an 85 dB reference level.

Perfect settings may be unattainable, but if you must err, do so on the plus side for the front-center channel and on the minus side for the rear speakers. You want the front-center speaker to deliver intelligible dialogue; you don't want the surrounds to call attention to themselves.

Repeat this process each time you add new speakers to the system, or even when you just move a single speaker.

If you are not relying on auto setup for surround levels, you may also need to manually key in speaker distances.

Other surround settings

Even in receivers without auto setup, setting **delay times** has become easier. Getting delays right used to require the user to calculate and choose the correct number of milliseconds based on positioning of speakers and listener. Nowadays, rather than make you struggle to understand what the settings mean, a receiver will simply ask you how far you're sitting from each speaker, and make the appropriate choices based on your responses. Key in the distances and you're done—no muss, no fuss, no mathematics. Auto setup does this for you. Some receivers allow advanced tweakers to adjust delay manually.

Setting the surround delay time adjusts the soundfield between the front and rear channels to ensure that ostensibly simultaneous sounds from each speaker arrive at your ears at about the same time. In Dolby Pro Logic mode, where sound tends to leak between channels, there's a second reason for the delay—it minimizes **Haas effect**, the perception that whenever sound arrives from two sources, you perceive it as coming from the closer source. Delay ensures that, even if you're sitting physically closer to the surround speakers, sounds that are supposed to come from the front will sound that way.

Bass management settings

Bass management settings direct low bass frequencies—the hardest ones to reproduce—to the speakers best suited to handle them. Auto setup will handle this. If you'd prefer to second-guess auto setup, tell the receiver whether each of your speakers is **large** (with full-range woofers) or **small** (monitor or satellite speakers that need a subwoofer for bass reinforcement). The size of the speakers is really less of an issue than their frequency response; a "large" speaker is one that reproduces frequencies below the crossover frequency, usually 80 Hz. You'll be asked whether the system has a subwoofer (a yes/no question).

One common setup would be "large" front left and right speakers and "small" center and surround speakers. If the main speakers provide enough bass to eliminate the need for a sub, the subwoofer channel is shut down altogether ("off") and bass frequencies are routed to the main speakers. (The THX people, among others, encourage you not to run "large" speakers with a sub. Some THX gear will prevent you from doing this, though in most cases it's more a suggestion than a requirement.)

Your receiver may be able to mix bass from the front left and right channels into the subwoofer output. This is a way of second-guessing the mixing engineer. Try it, if available, and see if you prefer it.

If your receiver has power to spare—which is rarely the case—and you're doing without a sub, you may want to run all of your speakers full-range. Simply set the sub to "off" and everything else to "large." You'll get the soundtrack's full bass response all around. (Unfortunately, many soundtrack mixing engineers elect to omit low bass frequencies from the surround channels under the assumption that most systems won't have full-range speakers in the rear.) At the opposite extreme, systems that use small satellites all around cannot function without a subwoofer, and with these systems the receiver must send *all* bass to the sub. Set speakers to "small," subwoofer "on."

With powered towers, the main speakers' side-mounted active drivers effectively function as subwoofers. The built-in subs may receive either direct input from the receiver's subwoofer output, if they have line-level inputs (speakers "small," sub "on"), or may derive the subwoofer signal from the main left/right speaker connections (speakers "large," sub "off"). Check the speaker manual.

Adjust subwoofer controls

Nearly all subwoofers come with a few standard controls. The **volume** control, of course, adjusts the sub's internal amplifier relative to the rest of the sound (so does the receiver's setup menu). You can try using the sound meter to measure the sub's output (be sure it's set to C-weighting to include bass frequencies). But you'll probably get wildly different readings in different parts of the room. If you're like most people, you'll start out by setting the sub volume too high because, after all, you paid for the thing and you want to hear it. But over time, you'll discover that subwoofers work best when they disappear into the sound-field, providing only subtle reinforcement most of the time, and playing loudly only at peak moments. Feel free to experiment; I predict you'll end up with a lower setting than you started with.

The **crossover** control determines the frequency at which the sub takes over bass-producing duties from the satellite speakers. Read the specs for your satellite speakers to find out where their bass response begins rolling off. The speaker manual may suggest the right crossover. Set the crossover too low and you'll end up with a **notch** (or dip) in bass response. Set it too high and you may hear voices booming out of the sub, though high crossovers are a must with small satellite speakers. Room acoustic conditions may affect the crossover setting.

Most systems will have subwoofer crossover controls both in the surround processor (receiver or pre-pro) and in the sub itself. If the signal passes through both crossovers, an undesirable effect called **cascading** will result. In the best of all possible worlds the subwoofer would have a bypass switch (or a separate **LFE input** that bypasses the crossover) and you'd set the crossover in the surround processor's menus. However, if the sub doesn't have a bypass, and the signal must pass through both crossovers, open one of them up all the way by setting it as high as possible. That will minimize the cascading effect.

Which crossover is better? That may spend on the **slope** of the satellite speakers—in other words, how steeply they roll off bass frequencies. Most crossover controls don't adjust slope. If your sub and sats come from the same manufacturer, the sub crossover's slope may be more appropriate than the one in your surround processor.

The **phase control** helps smooth out differences in bass response that occur when the subwoofer is a different distance from the listener than the main speakers—as it often will be, for either performance or aesthetic reasons. You want all sounds to reach your ears at the same time. Adjusting the phase control determines the direction in which the sub's drivers move, which in turn affects how bass is propagated from the sub. The phase control may be continuously variable, but usually has just two settings, 0 and 180. Experimentation will determine which phase setting produces better bass.

Program remote

Depending on what kind of remote you've chosen, programming it may take some doggedness and dedication. A preprogrammed remote requires just a few steps to identify your components to the remote. Getting a learning remote to do everything you want will take longer. In this respect, web-setup USB remotes are superior.

The inner workings of remotes are variable enough to make it impossible to give step-by-step instructions here. Yes, you'll have to read the instruction manual, even if you're a man, and program the remote with the manual by your side. But here are a few tips to get you started.

Remote capacities vary, both in the number of components handled, and the number of functions for each component. Don't drive yourself crazy trying to get your learning remote to do everything—there are times when it's easier to use front-panel controls or the original remote. Instead, prioritize. Concentrate on getting the universal remote to handle the most regularly used functions.

Once you've got the basics under your belt, you can start taking advantage of macros, which shoot multiple commands at the press of one or two buttons. That can be useful because there are sequences of commands that come up over and over. For instance, when you're getting set to watch a Blu-ray disc, you'll power up your DTV, receiver, and player, select the BD input on the receiver, and eject the drawer to receive the disc. Such repeated command sequences are prime candidates for automating via macro commands.

Macro commands are also a great way to organize multiple rounds of switching in a complex system. If some video sources connect to the receiver, and others directly to the display, you shouldn't have to remember that every time you switch from one source to the other—make the remote do it. If video and audio require dual switching for some sources, let the remote do that too. A little time invested in programming macros today will save you a lot of embarrassment in front of guests tomorrow.

Acoustic tweaking

Tweaks—defined as small things you can do to produce minor improvements in system performance—can be **acoustic** or **mechanical**. Of the two, acoustic tweaks are by far the most effective.

The room is not just an empty vessel to be filled with people, technology, movies, and music—it's another component in the system. Reproduced sound is a combination of **direct** and **reflected** sound. In a concert hall, the majority of the sound reaching your ears is reflected. In a home theater, it's better to cut down some of that reflected sound, and try to get more direct sound.

There's an entire industry devoted to acoustic tweaks based on advanced (and costly) technology. However, you can improve the sound of your space without hiring an acoustician by doing a few simple things.

The most damaging reflections are those from the floor and the side walls. Reflections from these areas make the sound abrasive and incoherent (not exactly a winning combination). So controlling reflections from these areas will help prevent room acoustics from spoiling the surround sound presentation, which is so critical to the suspension of disbelief that gives the home theater experience its emotional power.

First, to stop sound from reflecting off the floor, get a rug. Cover all or most of the floor with it. And make sure there's a carpet pad beneath it so that the end result is extremely thick. (I keep an old shag rug under my oriental rug; it's like walking on a cloud, and is acoustically absorptive.) The most famous practitioner of the rug strategy is President Jimmy Carter, who tells a story about the time he invited Vladimir Horowitz to play at the White House. The pianist listened to the acoustics of the East Room and refused to perform until the president had gone upstairs to fetch a carpet. Carter, the perfect host, obligingly moved the rug around the room until Horowitz was satisfied with the acoustics. Floor damping will benefit Republicans and Democrats alike.

Next, to stop sound from reflecting off the side walls, cover the side walls just in front of the main speakers to the extent possible. An acoustic treatment doesn't have to look like an acoustic treatment— thick drapes or shelves stuffed with books or disc media can work beautifully. Otherwise, consider tapestries, wall-hanging rugs, egg carton (whose variable surface diffuses side-wall reflections), or a thick sheet of foam rubber. Even an object just one or two feet square can make a big difference in taming side-wall reflections.

If the seating is right up against the wall—as sofas often tend to be in midsized or smaller rooms—treat the wall behind the listening position with something that diffuses sound. Otherwise the wall will reflect too much bass, and muddy everything else.

How do you know when to stop deadening the room? While a 5.1-channel home theater system requires more damping than a 2-channel music-only system, it is possible to go too far. A good way to get a quick fix on room acoustics is the **slap echo test**. Just clap your hands and listen to the sound as it bounces off the walls and decays. If it decays quickly, with very little reverberation after the initial slap, the room is in good shape. If you hear several distinct sounds a fraction of a second after the original, you may have work to do, and that can be fairly obvious—your own voice will sound echoey too. Don't over-do it. You'll want to have a little room reverb left to make music sound right.

Finally, pay attention to how your furniture affects the dispersion of sound. Avoid high seat backs that prevent sound from the side- and back-surround speakers from reaching your ears. Your system won't have a chance to sound good if the chair you're sitting in undermines it.

Mechanical tweaking

Killing mechanical resonance is an audiophile obsession that has inspired a whole genre of accessories that do one of two things. Either they dampen mechanical energy, by transforming it into tiny amounts of heat, or they simply move it elsewhere. These are controversial tweaks. Not everyone agrees that they work, or how they work.

Energy dampers often look like hockey pucks—you'll see them serving as the feet of certain components. Placed under components, they absorb vibration that otherwise would pollute delicate circuits.

Energy couplers, in contrast, are usually steel cones or spikes that soak up energy on their broad sides and transmit it through their sharp tips. Depending on whom you ask, their primary benefit is either to drain mechanical energy from components, or to prevent floor-borne vibration from affecting components. They may be used for just about any component (a disc player, an amplifier, even a rack) but are most often found in floorstanding loudspeakers and speaker stands, which often have threaded inserts on bottom to accept spikes. You might enjoy the physical sensation of low bass being transmitted through the floor and up through your seat. But the real reason to spike your speakers is remove extraneous music-polluting energy from their enclosures. Spikes do an especially good job of quelling the riot of vibration in a subwoofer enclosure. A spiked sub produces cleaner bass pitches.

The upgrade itch

Today an average lifetime encompasses several generations of changing technology. Therefore it's not impractical to consider home theater system upgrades every few years. With HDTV well entrenched and UHDTV on the rise, video display technologies continue to evolve. Lossless technologies are starting to transform surround sound and computer audio for the better. Internet-driven streaming, control apps for mobile devices, and USB DACs are the next wave. A home theater system in step with the times is open to change.

There are intelligent upgrades and there are stupid upgrades. In the latter category are inferior UHD, HD, and sub-HD flat panels, muscle amps far more powerful than smallish speakers and rooms require, a set of speakers that resembles Stonehenge and hogs the room, and cables that cost more than major components (though adequate cable is a necessity). At the other extreme, the smallest satellite speakers and cheapest receivers are subject to the laws of physics and economics, and to hope otherwise is wishful thinking. As Lancelot Braithwaite observes: "Bigger is not necessarily better. Smaller is not necessarily better." And while we're at it, even flatter is not necessarily better.

Upgrades that are worth it improve your system's performance in a manner that average eyes and ears can appreciate. That includes the quality of video and audio but there's more to it than that. For home theater to be a viable lifelong passion, it must fit into a practical life—or several lives if you have a spouse, family, or roommates.

So don't just think of ways to make your home theater system bigger or smaller. Don't feel obligated to chase the latest fad format (unless you're likely to collect a substantial library in that format). Rather, think of ways to make your system more elegant, stealthier, easier for other household members to use, more … practical. Replacing a big butt-ugly analog television set with a high-def flat panel or projector, or replacing the huge speakers you bought in your foolish youth with smaller ones that sound better—these are the smart ways to upgrade a home theater system. The job, if you love home theater, is never really over.

Connection glossary

Following, in numerical and alphabetical order, is a list of connections and associated terms used in home theater systems.

1/8-inch mini-jack/plug: Though rare in home theater systems, these space-saving connectors are common in PCs and portable devices. You'll need a mini-to-RCA adapter to connect a PC's soundcard or portable device's headphone jack to your system. Most mini-plugs are 1/8-inch (3.5mm) but some phones use a smaller 2.5mm version.

1/4-inch headphone jack/plug: The standard in home-audio headphone connections (originally known as the *phone* jack/plug and used by early 20th century Bell System operators).

5-way binding posts: Screw-down speaker terminals that accept spade lugs, banana plugs, double banana plugs, pin plugs, or bare wire. *Collared* binding posts do not accept spade lugs.

5.1-channel to 7.1-channel inputs/outputs: Six to eight analog RCA-type jacks provided to connect SACD, DVD-Audio, and some Blu-ray and DVD players. Also used to connect a surround preamp-processor to a multichannel power amp.

12-volt trigger: Used to send control signals between components, usually to coordinate startup.

75 Ohm antenna input: The type used on most TVs. Converts to 300 Ohms with an adapter.

300 Ohm antenna input: The type used on antique TVs. Converts to 75 Ohms with an adapter.

A/B speaker terminals: Some receivers allow connection of a second stereo pair.

Analog: A signal carried as a continuous waveform (as opposed to a *digital* signal, which is carried in pulses representing zeroes and ones).

Apple connector: Usually refers to the 30-pin connector used with Apple products. Recently succeeded by the smaller *Lightning connector.*

aptX: An audio compression codec with many uses. Most likely to affect an a/v system as a way of improving Bluetooth audio quality.

ATSC: The standard-setting body that originated the DTV standard, the

Advanced Television Systems Committee, also gave the standard its cryptic acronym of a name. ATSC 1.0 is the digital TV standard in the U.S., Canada, and Mexico; *NTSC* was the analog TV standard.

Audio: Besides the word's generic use, usually refers to a pair of RCA-type analog line-level audio connectors, but could include digital types.

Audio/video: Besides the generic use, generally refers to a pair of RCA-type analog stereo line-level audio connectors (color-coded white and red), usually mated with a one or more video connections.

AWG: American Wire Gauge, a measure of the thickness of cable. In home theater it applies mainly to speaker cable. An average gauge is 16. The gauge number drops as thickness increases, so 14 AWG is thicker than 16 AWG, and 18 AWG is thinner. Gauges higher than 18 or lower than 12 are not advisable.

Banana plugs: Inch-long plugs that easily connect speaker cables to speakers and amps equipped with binding posts. The flexing type provides the best fit—other kinds tend to fall out of the post. May be built into *double banana-plug adapters*, which are helpful for converting bare wire tips to bananas.

Binding posts: Speaker terminals that accept a variety of specially terminated cables. Opposite of *wire clips*. *Five-way* binding posts accept spade lugs; UL-approved *collared* binding posts do not.

Biamp/biwire speaker terminals: Allow connection of two sets of speaker cable. Biamplification adds extra amp channels to provide speakers with more power. Biwiring gives each speaker driver a direct path to the amplifier; its benefits are more subtle.

Bluetooth: Wireless protocol used to connect mobile devices to a/v systems. It is a direct device-to-device connection that does not rely on a home network. May be used with various kinds of compression coding (aptX, AAC) for higher-quality transmission.

CEC: Consumer Electronics Control. It enables components connected via HDMI 1.2a (and up) to coordinate commands. Though it is a non-proprietary standard, CEC is often disguised with proprietary names

Above are some jacks you're likely to see on the back panels of various components. From top and left: wire clips (PSB), 5-way binding posts (JBL), biwire binding posts (Platinum Audio), multi-use RCA-type inputs (Pinnacle), Toslink digital optical inputs (both uncapped and capped, McIntosh), XLR balanced analog line input (McIntosh), S-video jacks (Zenith), RF terminals (Zenith). In the lower righthand corner, from top, are IEEE 1394, HDMI, and DVI jacks.

223

such as Panasonic's Viera Link, Sony's Bravia Theatre Sync, Sharp's Aquos Link, Toshiba's Regza Link, and others.

Characteristic impedance: A laboratory measurement that measures a cable's resistance to current while assuming that the cable is of infinite length. That's impossible, of course, but the aim is to specify the characteristics of the cable itself and not what it's connected to.

Collared binding posts: Speaker terminals that accept banana plugs, pins, and bare wire but not spade lugs. *Five-way* binding posts do accept spade lugs.

Coaxial (vs. optical): Usually refers to a digital audio connector using an RCA plug or jack. Also frequently used to describe the 75 Ohm RF-type cable used to deliver cable TV. More generally, refers to any cable with an outer sheath surrounding an inner conductor.

Component video: A direct analog HD-capable video connector that separates three different parts of the video signal, a brightness signal and two color-difference signals, usually with RCA-type connectors, color-coded red, green, and blue. Better than S-video, composite video, or RF-modulated video signals; not as good as HDMI or DVI. May be labeled *Y Pb Pr* , *Y Cb Cr*, *Y B-Y R-Y*, or *YUV*. Used to connect DTVs to set-top boxes, DVD players, and other signal sources. Being phased out due to Hollywood copyright concerns.

Composite video: An analog video connection using an RCA-type connector, usually color-coded yellow. Not HD-capable.

DBA-25: Multi-pin connector that carries 5.1-channel analog line-level signals. Invented by THX. Rarely used.

Digital: A signal carried in pulses representing zeroes and ones (as opposed to an *analog* signal, which is carried in a continuous waveform).

DTCP: Digital Transmission Copyright Protection, a copy-prevention scheme, supported by several TV makers, that prevents unauthorized copying through the IEEE 1394 digital interface. One of two copy-prevention schemes used in a few DTV displays, set-top boxes, and recorders. *DTV Link* is an IEEE 1394 interface with DTCP.

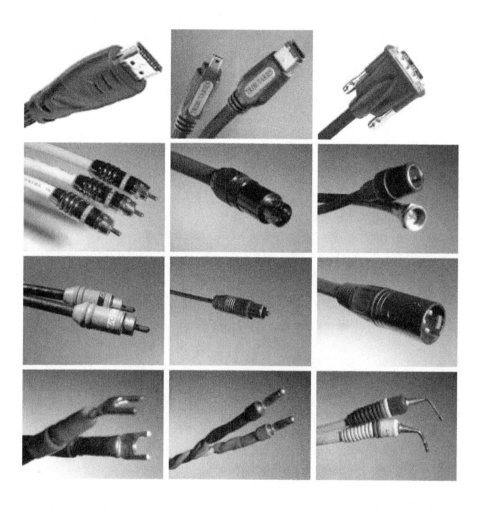

Here are some plugs and other termination hardware often found on the tips of inter-connect and speaker cables. From top and left: HDMI (Monster), IEEE 1394 (Tributaries), DVI (Monster), component video (Tributaries), S-video (Tributaries), RF coaxial (slip-on type at top, screw-on type at bottom), the multi-use RCA plug (XLO), Toslink digital optical, XLR (Esoteric), spade-lug speaker cable (MIT), banana-plug speaker cable (Straight Wire), pin-plug speaker cable (Monster).

225

DTV Link: See *IEEE 1394.*

DVI-D: Digital Video Interface, created by the computer industry's Digital Display Working Group, to handle digital signals. The DVI signal is not uncompressed, contrary to popular opinion, though it does have huge amounts of peripheral data deliberately added to make it too unwieldy for digital recording or networking. DVI signals are no better or worse in quality than component video signals. Other versions include DVI-A, for analog signals, and DVI-I, for both digital and analog signals. An adapter can *physically* mate DVI to HDMI but the two connected devices will work together only if they share the same signal protocols, chiefly copyright protection.

DVI-HDCP: Digital Video Interface with High-bandwidth Digital Content Protection. The latter is heavily supported by Hollywood, and is one of two copy-prevention schemes used in an increasing number of DTVs and set-top boxes. Now superseded by *HDMI,* which supports audio as well as video. DVI converts to HDMI easily with an adapter.

F-connector: Plug found at the tips of the RF cables used for cable TV and other video delivery sources.

FireWire: See *IEEE 1394.*

Fixed output: An audio output whose signal cannot be adjusted by the volume control. The opposite of *variable.*

Front-panel inputs: Jacks located on front of TV or other product, convenient for connecting mobile devices, games, or camcorders.

HDBaseT: Interface that carries HD and UHD video, audio, and power up to 328 feet using generic Cat5e or Cat6 cable with RJ-45 connectors. This new interface is especially useful for projectors, which need long cable runs, and multi-zone applications.

HDCP: High-bandwidth Digital Content Protection, the kind used with HDMI and DVI. Version 2.2 is the latest and UHD-capable version.

HDMI: High-Definition Multimedia Interface, a new combination of *DVI* plus multichannel audio, stereo audio, integrated remote control, infrared repeater, and IP signals. Despite its all-embracing nature, the connector is actually smaller than the DVI connector, and has untangled much of the rat's nest of cables that plagues home theater systems. The original *HDMI 1.0*, released in 2002, supports video and stereo audio, not surround. *HDMI 1.1* adds Dolby Digital Plus, Dolby TrueHD, DTS-HD High Resolution Audio, and DVD-Audio. *HDMI 1.2* adds Super Audio CD. *HDMI 1.2a* adds CEC. *HDMI 1.3* came out in 2006 and supports DTS-HD Master Audio, higher bandwidth, greater color depth, a new mini-connector for camcorders, and automatic lip-sync to keep soundtracks coordinated with faces speaking onscreen. *HDMI 1.3a* and *1.3b* make small modifications relating to compliance testing but offer the consumer no functional improvements over HDMI 1.3. *HDMI 1.4* adds 3D, IP device sharing, data exchange, other features. *HDMI 1.4a* has a fuller set of 3D options for movies, broadcasts, and games. HDMI 2.0, adopted in 2013, raises the frame rate for UHD video from 30 to 60 frames per second, doubles bandwidth to 18 gigabits per second, supports the 21:9 aspect ratio, delivers dual 1080p video streams to one screen, supports up to 32 discrete audio channels with audio sampling rates up to 1536 kilohertz, and is backward compatible. HDMI 2.0a, adopted in 2015, adds metadata for the HDR variant of UHD.

IEEE 1394: Generic designation for a high-speed digital connector available in four- and six-pin versions. The four-pin version is more common in a/v products, the six-pin version in computer products. Also known under such proprietary names as Apple's *FireWire* and Sony's *iLink*. Named for the Institute of Electrical and Electronics Engineers, which originated it, plus the document that contains its specs. *DTV Link* is an IEEE 1394 interface with DTCP copy-prevention features. It is rarely used in home theater products compared to HDMI.

iLink: See *IEEE 1394*.

Impedance: The load that a speaker presents to the amplifier. *Nominal impedance* is the specified number, though in practice impedance varies with the signal, and routinely drops below that number. *Characteristic impedance* relates to cables.

LFE: Low frequency effects channel. It carries bass information from a surround receiver or pre-pro to a subwoofer.

227

Lightning connector: Heir to the 30-pin connector for Apple devices.

Line level: The standard range of signal voltages used by receivers, disc players, and many other home theater system components.

MHL: Mobile High-Definition Link, a new audio/video interface designed to patch a smartphone into an HDTV or a/v receiver. It has a compact 5- to 11-pin plug at one end (usually, but not limited to, mini-USB) and an HDMI plug at the other. It carries 1080p video and 7.1-channel audio, both uncompressed, and allows the TV to charge the mobile device.

Mini-jack/mini-plug: The 1/8-inch jack and connector used in most portable devices.

Monitor output: On surround receiver or preamp-processor, an HDMI, component video, S-video, or composite video output feeding a video display.

MTS: Multichannel Television Sound, the official name of the standard for stereo analog television broadcasting, invented by Zenith. Usually includes dbx noise reduction and always accompanies SAP (Second Audio Program). May survive in analog cable channels.

Multi-room: Audio system serving more than one room. Also called *multi-zone.*

Multi-room/multi-source: Audio system serving more than one room with more than one simultaneous audio or a/v feed.

Nominal impedance: Speaker specification that describes a speaker's resistance to current and therefore the load that it presents to an amp.

NTSC: The North American television standard, named for the standard-setting body that originated it, the National Television System Committee. Adopted in 1941 and updated for color in the early 1950s, NTSC was the analog TV standard in the U.S., Canada, Mexico, and Japan; the U.S. discontinued it in June 2009. Replacing it is *ATSC 1.0*, the digital TV standard in the U.S., Canada, and Mexico.

Optical (vs. coaxial): Usually refers to a digital audio connector using fiber-optic cable. Some believe digital *coaxial* connections to be superior.

PCM: Pulse code modulation, a generic type of digital audio. Uses and resolution vary. For instance, a Blu-ray player may convert lossless surround to a high-res PCM signal for a receiver that lacks DTS-HD Master Audio or Dolby TrueHD decoding.

Phono input: On a receiver or preamp, an input that accepts the fragile low-level audio signal coming from a turntable and phono cartridge. Boosts and equalizes the signal for listening on a home audio system.

Pin plug: A slender form of speaker-cable termination that mates with wire-clip speaker terminals.

RCA (vs. XLR): Connectors using the quarter-inch-wide plug and jack invented by RCA Laboratories. Sometimes referred to as *unbalanced* (nothing wrong with that, just a technical term). RCA jacks are widely used in audio and a/v gear for analog audio, digital audio, composite video, and component video connections.

RF (radio frequency): In television, a 75 Ohm connection that carries multiple channels of video and audio, often used for antenna, cable, or satellite connections. In remote controls, a signal more robust and with greater range than infrared.

RF-modulated digital input: In home theater, the rare special input required for a receiver or preamp-processor to accept Dolby Digital signals from a laserdisc player.

RGB+HV: A form of analog component video found in high-end, pro, and computer gear. Includes red, green, blue color signals plus horizontal and vertical sync.

RS-232: A computer-type port used in custom installation to coordinate operation of touchscreen interfaces, firmware upgrades, other functions.

SAP: Second Audio Program, the ability to switch a TV broadcast from one soundtrack to another. For bilingual broadcasts among other uses.

Spade lugs: Y-shaped speaker connectors that mate only with non-collared binding posts (they will not fit wire clips).

S/PDIF: Sony/Philips Digital InterFace, the digital audio connection used in receivers and other digital audio/video products. May be *optical* or *coaxial* types.

Speaker level: Connection using speaker cables, usually referred to in connection with subwoofers.

Stereo: Two-channel signal. One mode of many on a surround receiver. The way most music has been recorded for the last half-century. An aesthetic that brings together low-end compact-system users and high-end audiophiles. In its original form (now archaic), the word *stereophonic* referred to anything that provided a three-dimensional sonic perspective.

Subwoofer: A speaker, usually with internal amplification, that produces only low bass. A receiver's or pre-pro's subwoofer or *LFE* (low frequency effects) channel carries the low bass signal to the subwoofer.

S-video: An analog video connection that splits the signal into brightness and color. May be labeled *Y/C* (Y is brightness, or luminance, while C is color, or chrominance). Better than composite video but not HD-capable. Recently eliminated from most receivers and other products.

Toslink: Official name of digital optical cable with plastic (as opposed to glass) filament. Toslink jacks are found on receivers, disc players, set top boxes, streamers, servers, and other source components.

Variable output: An audio output adjustable through the volume control. The opposite of *fixed*.

VGA: 15-pin computer monitor connector.

Video: When used without qualification, the term usually refers to composite video connections, but could also refer to other types.

Wire clips (spring-loaded or clamp-down): Speaker terminals designed to receive bare wire, usually 16-gauge or less, and not much else. Universally considered inferior to *binding posts*.

Y/C: A video connection that separates *Y* (luminance/brightness) from *C* (chrominance/color). See *S-video*.

XLR (vs. RCA): 3/4-inch-thick connector using a three-pin jack or plug. Carries line-level audio signals for long distances, either in professional installations or high-end home stereo systems. Rare in surround gear. Also referred to as *AES-EBU*, having been adopted by both the Audio Engineering Society and the European Broadcasting Union. Sometimes referred to as *balanced*.

Problem solving

You can find instruction manuals for many products on manufacturer websites, often in Adobe Acrobat format (the Acrobat Reader is a free download at adobe.com). Here—loosely based on the "troubleshooting" guides often found in those manuals—are ways to solve some of the problems that arise in home theater systems. What kind of problems? Well, problems with...

TV sets: basic issues

No power: Is the TV powered up and plugged in? Just asking! This may seem insultingly obvious, but according to tech support people, it's one of the most common problems they hear about. I've refrained from re-peating it for other products, though it applies widely.

No picture or sound: Is the source component on? Are you using correct input on TV and/or receiver? Check physical connections. Check input menu settings on TV and receiver, and output settings on source ensure signal is getting through. When not used with receiver, TV needs separate antenna, cable, satellite, or network connection.

No picture: When routing video through receiver, use same video-connection type throughout signal chain (unless receiver converts all incoming signals to HDMI, as most do now).

No sound: Check audio menu settings on TV, receiver, and source component to ensure that signal is getting through. Note that component video, S-video, and composite video connections do not carry audio (only HDMI and RF carry both video and audio). When viewing DTS DVD, turn on DTS output in player menu.

Cannot tune broadcast or cable channels, or some channels have disappeared: Run (or rerun) channel scan. Note that more cable systems are applying decryption even to basic broadcast channels, requiring a set top box or CableCARD.

Frozen picture, intermittent sound, or distortion on broadcast channels: Check signal meter—antenna may be insufficient or pointed the wrong way for some channels.

No internet-related features: Connect ethernet or (if available) wi-fi to network. Input router password into component. Router's security scheme may not be supported by smart TV or may require a more challenging manual network setup. Configure media sharing in PC.

No response to remote: Aim remote directly at TV or component. Aim remote at a less extreme angle. Inspect batteries for correct polarity (+/-). Replace dead batteries.

TVs and projectors: picture-quality issues

Color is unnatural: Quickest solution is to pick TV's movie mode. For fine-tuning, adjust color or tint controls. Adjust "color temperature" or "color balance" to 6500 Kelvins ("movie mode" or "low"/"natural" setting). Consider having your TV calibrated by an ISF technician.

Dim picture: Plasmas and especially projectors need ambient-light control. Darken room, with blackout curtains if needed. Try projection screen with higher "gain" (reflects more light back to viewer). If problem is most severe at sides, a lower-gain screen will have a wider "sweet spot." Projector lamps quickly lose brightness early in lifespan. Again, dark room is best solution. Reduce size of front-projected picture.

Painfully bright picture: Most TVs are factory-set way too bright and contrasty. Turn these settings down. Projectors can benefit from a lower-gain screen.

Oddly shaped picture: Adjust aspect ratio (screen proportions) to suit programming: 16:9, 4:3, letterboxed, anamorphic, etc. This adjustment is found in video source components as well as DTV sets.

Blank bars at top/bottom of picture: Programming wider than a 16:9 set will use the bars to deliver full width. This is normal and preferable to a stretched picture.

Blank bars at sides of picture: Programming narrower than a 16:9 set (4:3 content) will use the bars to avoid stretching. You'll also see this with 16:9 programming on an ultra-widescreen set. This is normal and preferable to a stretched picture.

Bar rolling through picture: If it appears on all sources, TV may be inadequately grounded. Make sure outlet is grounded. If it appears on only one source, check source's grounding.

Make those dots go away! Increase viewing distance or reduce projected area. Fixed-pixel displays may have visible pixel grid. Cut detail.

Diagonal edges jagged, motion distorted, backgrounds shift, other weird stuff: Cheap line doublers and other inadequate video processors cause artifacts. Upgrade TV or Blu-ray player or add scaler. Up- or downconversion between video formats can cause unpredictable effects.

Blotches mar large areas of the screen: In an LCD TV, especially the LED-backlit type, uneven backlighting can cause uniformity problems. The only solution is to buy a better TV. Plasmas may be subject to "false contouring." Some digital signal sources (internet, satellite, certain DVD releases) have an inadequate amount of video data, causing artifacts.

Bright objects rapidly moving against dark backgrounds show multi-colored flicker or trailing: This problem is inherent in one-chip DLP projectors using a three- or four-segment color wheel. Six-segment color wheels are less subject to it.

Noisy image: Nearby electrical products may be causing interference. Check fluorescent lamps, hair dryers, air conditioners, etc. Use power-line conditioner to isolate TV from other things in (and outside) system.

Dark spot on picture: Plasma and LCD sets are both subject to missing pixels. This is a manufacturing defect. If spots are numerous enough, exchange set immediately.

Hey, this isn't HDTV! SDTVs and EDTVs are not HD. HDTV programming available only via broadcast, cable, satellite HD-DVR, D-VHS (not regular VHS, DVD, LD). Internet content labeled HD may be stretching the definition. Not all channels or programs are HD.

Whirring sound: Cooling fan (in a plasma).

Speakers

Getting your speakers to sound right has a lot to do with the mysterious arts of component matching, cable matching, and acoustic tweaking. This may take some work.

Speakers don't sound as good as they did in the store: Your room is acoustically different from showroom. Dealer may have used better amp (big difference) or better cable (slight difference). Most speakers require a few dozen hours of break-in to achieve best performance.

Sound is compressed. Amp runs hot or shuts down: Bad news—impedance mismatch may require change of speakers or amp.

Bass is weak. Sound seems hollow, disembodied: One or more speakers may be miswired "out of phase." Check connection at both speaker and amp ends (red to red, black to black).

Overall sound is OK except for bass—too weak or strong: Move speakers closer to walls, corners, floor for more bass. Move speakers away from walls, corners, floor to reduce bass. Sitting close to the back wall also boosts bass, sometimes excessively, so sit closer to speakers if needed. Do not block ports or vents in back of speakers—get them off shelves. Subwoofer positioning requires special care and experimentation. See "Installation guide/Subwoofer placement."

Sound is abrasive and muddy: Deaden room acoustics by adding absorptive or diffusive elements such as rugs, curtains, soft upholstered furniture, music/video shelving, and other elements to keep sounds from bouncing off hard surfaces, especially floors and side walls.

Sound is vague. Can't seem to get the lyrics: Position tweeters at eye/ear level. Toe in speakers toward listening position to increase proportion of direct sound. Smaller speakers usually produce clearer sound when mounted on stands a couple of feet or more from the wall.

A buzzing or rattling sound is audible: The metal baskets holding the drivers may be loose. Tighten all screws around the drivers (and be careful not to let the screwdriver slip and puncture a driver). Plastic parts (like the back panels of certain soundbars) may resonate. Or the unwanted vibration may occur elsewhere in the room. Check ceiling tiles, light fixtures, doors, windows, moldings, bric-a-brac, etc. Don't leave things sitting on top of speakers. To make this problem easier to diagnose, try the 20Hz-20kHz "sweep tone" on many test discs.

Surround receivers & components

Here, at your system's heart, it's easy for a missed connection to foul up the works. Another common problem is signal routing—in other words, finding what button or menu will send video and audio signals where you want them to go. Correct setup is also vital: Your receiver must know how many speakers are attached to it, what channels are active, and other setup parameters. Connecting the supplied microphone and running the auto setup program will handle those details unless you're confident enough to do a manual setup.

Receiver shuts down: Power amps (including receivers) run hot, need ventilation space. If possible, move amp to top of rack, with nothing overhead. Never, ever put another component on top of the receiver. Leave at least three inches of space above the amp and an inch or two at the sides. Mismatched speakers may overtax receiver—check their sensitivity/efficiency ratings. You may need a more powerful receiver or more efficient speakers.

No picture: See is TV is on. See if receiver is set to desired source. See if TV is set to appropriate HDMI or video input. Check video connection from source to receiver. Check video connection from receiver to TV. Check video setting in receiver's input menu. Match video connection from source to receiver, receiver to TV (if receiver does not convert all outgoing signals to HDMI).

No sound: Check receiver volume or mute. See if receiver is set to desired source. Check audio connection from source to receiver. Check audio connection from receiver to speakers. Check audio setting in receiver's input menu. When playing BD or DVD, select appropriate audio setting from output menu. When playing any surround format, make it is supported. Protection circuit may be activated. Stop playback, lower volume, and power down receiver to reset it.

No sound in front left/right speakers: Select correct zone. Select speaker set A or B (A/B switch may be set "off" at factory).

No sound in front center and rear left/right speakers: Stereo source—activate a surround mode such as Dolby Pro Logic II music mode. When playing two-channel Dolby Digital mixdown on BD or DVD, set receiver to Dolby Pro Logic II cinema mode. If surround-encoded material does not play in surround, check whether center and surround channels are turned on in surround processor. If center or surround channels are too low, calibrate surround processor for center and rear channel levels.

Unwanted sound from surround or center channels: When playing stereo source material, if two channels are all you want, shut down surround modes such as Dolby Surround, Dolby Pro Logic II, DTS Neo:6, etc. The "pure" or "direct" mode will do this and eliminate all processing including room correction as well as the subwoofer channel. If you want two channels plus sub, the correct mode is usually "stereo."

A loud hum is audible: Check speaker cables for crossed wires. Unplug components one by one to find source of hum; use power-line conditioner to isolate it.

Distorted/weak sound, receiver runs hot on all sources: Speakers present too big a load to receiver. Get a higher-powered receiver or more sensitive/efficient speakers.

Sound is hollow, disembodied, or lacking bass: Phase is reversed—match speaker cables red to red, black to black.

One channel weak or intermittent: Check audio cables for damage or loose connection. Cassette deck heads need cleaning.

FM reception noisy: Move antenna to different position. Try longer/shorter antenna. Switch FM to mono to reduce noise.

DLNA media sharing does not work: Connect ethernet or wi-fi to network. Allow sharing in Windows Control Panel (7 and up) or Media Player for XP (Library tab/Media Sharing).

AirPlay does not work: Needs ethernet or wi-fi connection. Enable network standby in receiver. If not supported, install AirPort Express.

Bluetooth does not work: Pair device with receiver. May need to activate Bluetooth reception in receiver menu. Receiver may support Bluetooth only with proprietary or third-party adapter.

Blu-ray/DVD players

Usually a setup wizard will guide you through the necessary initial settings. You can always change them in the disc player's setup menu.

Disc does not play: Condensation may affect player or disc. Let stand 1-2 hours. Player may not be compatible with DVD-Audio, CD-R, CD-RW, other disc or file formats. Try another disc. Allow extra loading time for Blu-ray (it's slower than DVD).

Some buttons don't work: Some operations are disabled for some discs—for example, skipping or fast-forwarding through trailers and copyright warnings.

No picture and/or no sound: See if TV is on. See if receiver is set to BD or DVD input. See if TV is set to appropriate video input. Check connection from BD/DVD to receiver. Check connection from receiver to TV. Check connection from BD/DVD to TV (if they are connected directly). Use same video connection throughout signal chain—some receivers don't convert one video connection to another.

Distorted picture: Disc mastered with too much video compression. Disc for wrong region, does not match television standard.

Frozen picture or mistracking: Disc is scratched or dirty, often a problem with rental or library discs (or small kids, or adult slobs). Try cleaning with a soft cloth. Wipe gently from hole outward, not with circular motion.

No audio from digital output: Check digital connection between player and receiver. Enable that digital output from player output menu. Enable that digital input from receiver input menu.

No subtitles displayed: Turn on subtitles function in disc or streaming device menu. Most programming is captioned.

Audio (or subtitle) language not switchable: Disc may not have multiple languages or subtitles. May have to be enabled in main disc menu.

No BD-Live features: Player not connected to network.

No Bonus-View features: Disc does not have such features.

Picture not in correct proportions—objects appear distorted in shape: Try version of movie on other side of 2-sided disc. Select 4:3 or 16:9 from DTV menu. If using ancient analog TV, select "4:3 letterbox" mode from DVD-player menu. Some older DTVs automatically go into anamorphic mode with 480p input—again, try 4:3 letterbox mode (and suffer loss of resolution).

Satellite receivers

Since there are two competing satellite-TV formats, the generic advice below will not cover format-specific problems. Also consult the troubleshooting section in the manual, and the DirecTV (directv.com) or Dish Network (dishnetwork.com) websites.

No service or signal: System requires access card. Card must not be inserted upside-down or backwards.

No picture: Check signal strength. Use RG-6 (not RG-59) cable between dish and receiver. Remove signal splitters from signal path. Check for obstacles in signal path. Check for moisture in connections. Level dish antenna. Check elevation (up/down) and azimuth (side/side) positioning. Check connection between satellite receiver and TV. Check if TV or a/v receiver is set to correct input.

No picture or frozen, intermittent, messy picture: Inclement weather may cause "rainfade." Wet snow may accumulate in dish.

Not all channels displayed in program guide menu: Past or future programs cannot be displayed. You may be looking at a "favorites" list.

Oops, forgot the password: Call tech support with name, address, phone, serial number, PIN. If you're that absentminded, write it down, or use something easier to recall.

Antennas

There are two main ways to foul up a TV antenna. One, choose the wrong kind. Two, install it in the wrong place.

Signal is absent: Check all connections. Center conductor of RF cable may be bent to one side. If antenna is active (powered), make sure power is on. DTV (digital) signal strength may be too low for reception. Check all switching (such as video recorder's TV/VCR switch).

Signal is weak: Raise antenna to higher position to raise signal strength. Attic placement reduces signal strength; move antenna to roof. Aim antenna toward transmitter location (even if it's "omnidirectional"). Reposition antenna to minimize obstructions such as nearby buildings or towers. Remove signal splitters. Don't split more than needed. Don't use 6- or 8-way splitter when 2-way splitter will serve. Add amplifier (or use amplified antenna) to boost signal, especially with long cable runs. Larger antenna, or different type of antenna, may be needed. Be realistic about limits of transmission distance and terrain.

TV channels 2-6 are noisier than others: Minimize electrical interference (if possible) by moving antenna away from power lines. Look for small breaks in antenna cable (effects are more visible on lower channels). Try motorized antenna to shift antenna position as needed. These "lowband" channels may require a larger antenna.

Pixellated, artifact-ridden picture: Run the automatic channel scan from the TV's or tuner's setup menu to catch up with modifications made by broadcast stations.

UHF (analog or digital) channels are weak: Fold down UHF elements at base of outdoor TV antenna.

FM signal is weak and noisy: Switch from stereo to mono—mono signals are always stronger. This is a quirk of analog FM broadcasting.

DVRs, streamers, & servers

Devices that connect to an IP network or antenna/cable/satellite feed may have a variety of problems that can barely be summarized here. Here is just a smattering. (For TiVo owners, I recommend the "TiVo Repair and Troubleshooting Guide," available at dvrupgrade.com and fixmytivo.com.)

No picture and sound: Check the usual: video/audio connections, cables, inputs. Check network/cable/satellite status. Network password may be needed. Media sharing may need to be activated in computer. In Win 7 and later, see Control Panel/Sound. In Windows XP, check the Windows Media Player (Tools, Options, Library, Configure Sharing).

DVR or server—boot problems, no recording, freezing, stuttering, loud clicking: Bad news: defective hard drive. He's dead, Jim.

DVR shuts down: Those hard drives run hot. Is the fan working?

No program guide: Bad modem, or modem needs reset. Also, TiVo does not work with a VOIP line.

Low-resolution picture: Much streaming is just standard-def and even the high-def services suffer from substandard picture quality. Internet connection may be slow or clogged with other activities.

Turntables

A loud hum is audible: If using turntable, attach ground wire to ground terminal. Check speaker cables for crossed wires. Unplug components one by one to find source of hum; use power-line conditioner to isolate it.

Howling noise audible at high volume: Turntable/speaker feedback—move, isolate turntable. Or Kurt was really passionate that day.

Distorted/weak sound on LPs: Clean record and stylus before every play. Make sure phono preamp or phono input support cartridge (most surround receivers support only moving-magnet cartridges, not moving-coil). Match settings of stylus pressure and anti-skating.

Remote controls

Incorrect programming is the most common problem with remotes—before you assume the unit is broken, call the technical support line.

No response from components or remote indicator lights: Inspect batteries for correct polarity (+/-). Batteries too weak to program or use remote. Replace batteries. Button jammed. Remove batteries, press all buttons, put batteries back in.

Indicators light, but components don't respond to commands: Aim remote directly at component. Aim remote at a less extreme angle. Remote must be close enough to component (15 feet is common). Make sure remote is set to command the right component brand/model. Make sure remote is set to correct device mode (TV, BD, etc.). Make sure remote is set for use, not for changing components or learning commands. In macro mode, aim remote for full duration of macro. Component may lack ability to receive remote commands (check front panel for IR window).

Component responds to some commands, but not others: "Partial code"—remote is preprogrammed for a slightly different product. Look up correct model/code and reprogram remote.

Preprogrammed remote can't control a component: Hit all required commands before/after entering three-digit code. Check model number of component against list of supported models in remote manual. Older or rarer components may not be represented in remote's code library (try a learning remote). Even if a component is listed, not all of its functions may be programmable.

241

Learning remote will not accept new codes: Two or three tries may be needed to get remote to accept code. When transferring codes, use fresh batteries in both remotes. Inspect batteries for correct polarity (+/-) in both remotes. Align original remote and learning remotes at correct distance (see manual). Put a wedge under one remote to align it with the other. Use tabletop or other surface when transferring commands. Fire original remote for time specified by manual. Keep an eye on the indicators on both remotes (see manual). Make sure learning remote is in programming (not use) mode. Make sure learning remote is in right device mode (TV, BD, etc.). Hit all required commands before and after transferring codes. Learning remote holds fixed number of codes. Full? Eliminate nonessential codes. Other infrared devices in room may interfere. De-activate them. Direct sunlight or fluorescent lighting may interfere with transfer. In rare cases, some codes will not transfer. Use original remote.

Remote will not change TV channels: If original remote required "enter" command, hit "enter" on universal remote.

Power-line accessories

Problems here should be taken with an extra measure of seriousness. There may be issues concerning safety or your system's performance.

Equipment does not turn on: Check all connections. Switch on power-line accessory. Keep it switched on at all times to feed components operating in standby mode. Reset circuitbreaker, or change fuse, and do not exceed product's power-handling capabilities (see manual). Equipment may be damaged (see below).

System does not work—components fried: If your system was properly connected, get in touch with the surge-suppressor maker. You may be entitled to replacement gear plus a new surge suppressor.

Indicators unlit or lit with wrong color: Be certain power-line accessory is connected to 3-prong outlet. The third prong is "ground." You may use an adapter that attaches to the screw holding the outlet plate in place (the screw should be attached to ground). But do not cheat by using an adapter that is not connected to ground—a safety hazard or performance problem may result. This may be a warning that polarity is incorrect, or that ground is missing or improperly connected. These things could pose deadly hazards and must be taken seriously. Get a licensed electrician to check and repair defective outlet.

Cables

HDMI—no picture: HDMI not suitable for long runs. Use HDBaseT or component video or consult a custom installer for further solutions.

HDMI or USB—stuttering or harsh sound: These cables are subject to interference and jitter. Cables with ferrite cores may reduce interference. Audio components with jitter reduction (for example, an asynchronous USB DAC) sound better.

Optical cable—no sound: Cable will not conduct a signal if kinked. Replace cable.

Racks, stands, & mounts

Assembling them correctly is half the challenge. The other half is to correctly position your components for best performance.

Help, the parts don't fit together! Chill. Spread parts out on floor, with manual. Try to visualize how things fit together. If woodscrews or shelf hardware don't fit, loosen holes with an awl, slowly and gradually.

Cables behind your rack are a mess: Use rack's cable management feature. Use plastic twist-ties to attach cables to vertical support. Don't group power cords with other cables (it causes interference).

Metal rack flexes, is crooked: All-metal shelves from Boltz may flex if screws are excessively tightened.

Rack on wheels flexes when you move it: Always push or tug from bottom of rack, never from top. Move heaviest components to bottom shelves; don't stack components on top. Screw vertical supports together as tightly as possible. It may be unwise to move a heavily laden rack.

Newly relocated components act up: Keep them on separate shelves to allow proper ventilation, to minimize electromagnetic or radio-frequency interference, and to minimize mechanical resonances. Also keep power and interconnect cords as separate as you can. Relocate FM antenna to avoid interference from metal objects.

Newly relocated speakers don't sound as good: Reconsider new positioning. Fill hollow speaker stands with sand to reduce resonance. Attach spikes to base, coupling stand to floor.

Speaker wobbles on hinge, cannot aim in preferred direction: Tighten the screw that holds the hinge. Strengthen the hinge with a dab of glue. Best solution is mount rated for several times speaker's weight.

Manufacturer support

It happens in the best- and worst-run home theaters alike. Sooner or later, a component goes on the fritz. Then, in a scenario often more fraught with emotion than the works of Tennessee Williams, the hapless consumer must depend upon the kindness of strangers who work in manufacturer support departments.

You may well be about to spend a lot of time on the phone— preferably a headset phone, with some good music going in the background, assuming you still have a working audio rig. But this isn't the end of the world. And while you may feel helpless when you first realize something has gone wrong, there are ways to empower yourself, ways to make this little episode as painless as possible. Here, then, are some tips on how to survive a journey into the heart of tech support.

Be nice. This one comes first because it's the single most important thing about dealing with tech support people. No matter how upset you may be, it's important to realize that they are not responsible for the product going wrong (though you may well be). Nor are they responsible for the length of time you were kept on hold—they don't determine staffing levels, their bosses do. However, if you mind your manners, they can often (though not invariably) help make things right again. They are not your personal enemies. Their job is to help you, and they know it. Remember that—regardless of how long you were kept on hold. If you're calling a small high-end audio company, you might even find yourself speaking to the CEO!

Check the obvious stuff. Yeah, some of this may seem silly, but: Is the thing plugged in? If it's not responding to remote commands, are there batteries in the remote? (A cheap battery tester, available at any RadioShack, is an indispensable part of any home theater.) If it's winter and there's a lot of dry heat and static electricity in your home, have you tried powering down (or even unplugging) the unit to reset its frazzled electronic brain? Eliminate the obvious before making phone calls.

Read the manual before you call. Nearly all products come with manuals and nearly all manuals have a "troubleshooting" section. (So does this book—see the chapter on "Problem solving.") Troubleshooting guides will mention the obvious things above and a few things you may not have thought of. When you've eliminated those possibilities, give the rest of the manual a thorough read—especially if you've never done it before—to make sure you haven't simply misunderstood how the product works. Many of the calls tech support personnel receive are from people who haven't done their homework and just don't know which button to press.

Don't pop the lid. For heaven's sake, don't try to do surgery on the product yourself. If it's within warranty, your tampering will invalidate the warranty. Even if that's not the case, you could do more harm than good—perhaps to yourself. Electric shock is always a hazard even when the product has been unplugged. Its power supply and power capacitors, whose job is to store current, may still have electricity inside them. Unless you have either a degree in electrical engineering or a death wish, leave technology to the technicians.

Ask the dealer for help. If you've just bought the product, don't be a hero—exchange it. Even if that's not possible, your dealer may still be willing to provide support if the component is a high-end product. Specialty dealers get those big markups precisely because they provide better service than the mass marketers. High-end manufacturers often choose dealers who are willing and able to provide superior support.

Consider alternate modes of support. If the problem isn't urgent, get in touch with the manufacturer by email. You'll have to wait longer for a reply but at least you won't have to chew off your fingernails while being kept on hold. Also take a look at the manufacturer's website. In addition to the email address, the site may include other forms of support, such as FAQ (frequently asked questions), service-center contacts, online manuals (printable as replicas of the originals in Adobe Acrobat format), and other helpful stuff. Tech support links may be either on the homepage or tucked away in individual product listings along with specs and features. Manufacturer websites vary in scope and quality but it's never a bad idea to do a little homework.

Keep the number handy. Keep a paper file of all the manuals for all your home theater components and other appliances. The tech support number may be buried in the manual, or it may be printed on a small piece of paper enclosed with the product. Don't allow yourself to misplace that documentation—and certainly don't throw it away!

Know when to call. Most consumer electronics manufacturers do not provide 24-hour, 7-day customer service. Most provide support during business hours—which may be in the eastern, central, or western time zones—five days per week. Some have longer hours than others, though, and some are open weekends. The manufacturer's website may list the hours when support is available. If not, just call, but don't be surprised if you get a recording during non-business hours. At least the recording should tell you when to call back.

Know your model and serial numbers. One of the most common mistakes people make when calling tech support is not knowing the model number of the component they're calling about. DUH! Have the manual handy, and jot the serial number on the back page—you should be able to find it on the back panel or sometimes on the carton. By the way, keep the carton in storage for as long as possible, since you may need it someday to ship out a troubled component for service.

It's worth saying again—be nice! Customer service people do not have cushy jobs. They are under stress. They are not sitting in cozy private offices for the most part. They are generally in tiny cubicles, or large bullpens, wearing headset phones, dealing with cheesed-off or just plain confused people all day, surrounded by other tech support folks who are similarly stressed. Those who don't have a union to protect them may well wish they did (and I support them). If the person you're speaking to seems nice, be advised that it takes a huge effort. If not, just be glad that job isn't yours. If all goes well the tech support person may be able to help you. If you feel your problem was not dealt with in a reasonable manner, remember that the next time you consider buying another of that manufacturer's products. As a consumer you always have the right to take your business elsewhere.

Hiring a custom installer

While the main thrust of this book is to help the do-it-yourself home theater buff, some installations will require professional assistance.

Do you need an installer? Pulling cable behind walls isn't rocket science—a local electrician can probably do the job for you. (Be sure to choose fireproof cables.) On the other hand, if you need to install a projector mount, hang a screen, or map out a multi-zone audio system, you probably will need a custom installer. Anything involving a projector automatically requires a knowledgable technician to ceiling-mount the heavy object and adjust it for best performance. Poking holes in the wall to mount a heavy plasma or super-large LCD display—while minimizing violence to the wall—is one of many things an experienced installer can do better than an average civilian. Chain-store technicians can install satellite dishes and antennas but whether they install them well, to maximize signal strength, is another matter altogether. A good installer will also anticipate ease-of-use issues and set up your system so that any member of the family can operate it.

Consult trade groups for certification and referrals. Over the past few decades a whole industry has grown up around custom installation. Most of the better installers are certified by one or more organizations such as CEDIA (the Custom Electronic Design & Installation Association, cedia.org) or, for video, ISF (the Imaging Science Foundation, imagingscience.com). These organizations can also provide referrals to qualified custom installers in your area. Architects, interior designers, and cabinet makers also may be able to provide referrals.

Do your homework. To get off to a good start, make sure that the company is not only certified by at least one of the associations above but also fully insured and licensed to operate in the state where it does business. Speak to your local Better Business Bureau and state department of consumer protection to check whether a prospective installer has attracted complaints. Find out how long the individual or company has been in business. And, of course, obtain several references and check each one. Someone who has spent years installing car stereos or burglar alarms is not necessarily an expert on home theater or multi-zone audio. If the installer can't provide contact information for several projects comparable to yours, look elsewhere.

Verify manufacturer authorization. If you are interested in specific products or brands, make sure the installer has been authorized by the manufacturers. Manufacturer websites should point you toward authorized retailers. Don't settle for a dealer who equivocates—either he's authorized or he's not. If he's not you may end up with grey-market goods that lack proper warranties. The right installer/retailer is the one who provides superior service when a product goes awry, as even high-end products sometimes do. Smart manufacturers are scrupulous about who they allow (and do not allow) to sell and service their products.

Pay a visit. A good installer does not necessarily have to be associated with a large store but should present a businesslike appearance. "A visit to the installer's office is worth thousands of dollars," according to Mitchell Klein, president of Media Systems (Boston, West Palm Beach) and former president of CEDIA. Is the person neat and well-organized? Are your calls answered by a human being or at least returned promptly? This is who you will be calling when something goes wrong.

Talk, listen, and observe. The instant you begin speaking with a custom installer, information is being exchanged, and it moves in both directions. Of course you should write down your questions before the meeting and ask them. And your questions should receive satisfactory answers. But don't try to do all the talking. A good installer will ask about your goals, space, limitations, and budget. Having determined your needs and expectations, he will provide options and price quotes. Recognize that the cheapest quote may not result in the best service.

Get it in writing. When you're ready to proceed, get every detail of the job in writing. The proposal may include things like wiring, schematics, elevations, documentation, and project meetings. Try to anticipate cost overruns and upgrade issues. Unexpected problems may come up, especially in older homes that were not built with electronic systems in mind, though newer ones may have been pre-wired by the builder.

Be a good collaborator. Whether you need to be firm or flexible about arising problems is your own judgment call. If you've chosen a reputable installer, and find the person trustworthy, try not to be too hard-nosed. This is a relationship like any other. Treat a good installer as a collaborator and expected to be treated the same in return. A wisely chosen and intelligently handled custom installer should prove to be the best ally you could ever have imagined.

Notes & Acknowledgments

This is the 16th edition of an annually updated resource, revised in 2016 and dated 2017. Each edition generally runs from October to October of the next year. Please beware of retailers selling outdated editions. The best way to be certain you're getting the latest edition is to follow the links from quietriverpress.com to retail sites. This book is aimed at beginners and intermediate-level readers.

Home theater does not stand still. Its underlying technologies, big-screen television and surround sound, are rapidly evolving. *Practical Home Theater* will grow and change along with it. Please feel free to contribute comments and ideas for changes in future editions via email—see the contact page at quietriverpress.com. I may not be able to provide Q&A service but whatever you have to say will be read with interest.

Even before the first edition of this book came out in late 2001 I had planned annual updates. Over the years, parts of the book have been substantially reorganized and rewritten, especially the television and surround sound chapters. This year's edition includes additional material on the HDR format war, the next generation of cable readiness, the nascent ATSC 3.0 broadcast standard, and audio streaming players.

Some readers have asked why the page count has stayed the same for the past several print editions. The main reason is to avoid price hikes. Rest assured that every edition is reread, rethought, and revised.

That also means purging outdated material to make way for new or updated material. Indeed, some readers have complained about coverage of older technology. In response, I have downsized or eliminated much material that is less relevant (goodbye, VCR chapter). What you think of the book matters to me and I try to make it as fresh as you need it to be.

Many thanks to the expert readers who were kind enough to read and comment on portions of earlier drafts. Special thanks to Brent Butterworth (then of Dolby Laboratories, now contributor to *about.com*) for giving the audio coverage the benefit of his world-class expertise, and to Lancelot Braithwaite (former technical editor of *Video*, more recently with *Widescreen Review*) for helping me better understand the technology of television. They did the heaviest lifting and I can never thank them enough. Thanks to Jeff Samuels and Bill Schindler of Panasonic for helping with the 3DTV chapter—please note that, technical details aside, my skeptical take on the subject is entirely my own. I'm also grateful to Michael Guillory of Texas Instruments for help with DLP technology and to Scott Wilkinson for recommending Blu-ray test discs.

I couldn't have survived the chapter on "Understanding surround standards" without Craig Eggers (Dolby Labs) and Don Dixon (DTS). Thanks again to Craig Eggers (then of Toshiba) and the late Len Schneider (Technicom) for reading the chapter on DVD, and to Gregory Thagard (Warner Bros.) and Mikhail Tsinberg (Key Digital) for helping me better understand progressive-scan DVD. I'm also grateful to Brian Dietz and Neal Goldberg (NCTA), Bob Perry (then of Mitsubishi, now of Panasonic), John Taylor (LG/Zenith), and Mike Schwartz (CableLabs) for bringing me up to speed on digital cable readiness.

Joel Silver (Imaging Science Foundation) and Marty Zanfino (then of Mitsubishi) also shared their video expertise. John Dall (THX) and Graham McKenna (then of Pacifico PR) helped fine-tune the details about THX. Elliot Grimm of the Consumer Technology Association provided key statistics.

Please note that while these industry luminaries were kind enough to help weed out the garden, they and their employers or clients are in no way responsible for any remaining weeds. The author accepts blame for any errors. Opinions expressed are solely those of the author.

Thanks to the following folks for providing products photographed for the book: Joe Abrams (ex-Music Interface Technologies), Steven Hill (Straight Wire), the late Daniel Graham (Monster Cable), Joe Perfito (Tributaries), Stan Pinkwas (J.B. Stanton Communications), Phil Raymondo (Esoteric), Roger Skoff (XLO), and Bryan Stanton (J.B. Stanton Communications).

I'm deeply grateful to all of the editors who have supported my writing career with steadfast patience, kindness, and good editing. They currently include Rob Sabin and Claire Crowley of *Sound & Vision* (formerly *Home Theater*). I'll always be indebted to Doug Garr, my former boss at *Video Magazine*, for inspiring me to become a full-time writer. Like any working writer, I am indebted to more sources of collegial support than can be listed here.

Mark Fleischmann
New York, October 2016

About the Author

Mark Fleischmann is a New York-based writer specializing in technology and the arts. Currently serving as audio editor of *Sound & Vision*, formerly known as *Home Theater*, he reviews audio/video gear and writes news stories and blogs. He was a co-founder of *etown.com* (1995-2001), in its time the world's most widely read consumer electronics publication, and was its first and longest-serving editor-in-chief. Mark was audio critic of *Rolling Stone* for nine years and has written columns on audio/video hardware for *DigitalTrends.com*, *Premiere*, and *The Village Voice*. His column on home theater ran for 15 years in *Audio Video Interiors*, making it the longest-running column ever written on the subject. His writing has appeared in *Amtrak Express, Bloomberg Personal Finance, Business Week, Cargo, c|net, CrutchfieldAdvisor.com, Custom Retailer, Details, E-Gear, Elle Decor, Harper's Bazaar, Help.com, HX, The Men's Journal, Musician, Penthouse, Popular Science, The Robb Report, Spin, Stereo Review, The Stereophile Guide to Home Theater, Sync, Ultimate AV, The Washington Post,* and many other publications. Mark has served both as a movie/video critic (for *Entertainment Weekly, Newsday, Video*) and as a music critic (*Musician, Spin, Trouser Press*). A former senior editor of *Video* (1980-86), he edited the magazine's video programming section, as well as a record collector's magazine for *Trouser Press*. He is the author of another book, *Happy Pig's Hot 100 New York Restaurants*, converted to a website at *happypig100.com*. His other web destinations include *quietriverpress.com* (book publishing) and *mfwriter.com* (career survey). Mark still plays his LPs and still feels silly writing about himself in the third person.

Index

256